T0191750

Functional Safety for Road Vehicles

Hans-Leo Ross

Functional Safety for Road Vehicles

New Challenges and Solutions for E-mobility and Automated Driving

Hans-Leo Ross
Lorsch
Germany

ISBN 978-3-319-81494-0 ISBN 978-3-319-33361-8 (eBook)
DOI 10.1007/978-3-319-33361-8

Translation from the German language edition: *Funktionale Sicherheit im Automobil: ISO 26262, Systemengineering auf Basis eines Sicherheitslebenszyklus und bewährten Managementsystemen* by Hans-Leo Ross, © Carl Hanser Verlag GmbH & Co. KG.
Softcover reprint of the hardcover 1st edition 2016

Printed on acid-free paper

This Springer imprint is published by Springer Nature
The registered company is Springer International Publishing AG Switzerland

Foreword of the Author

The German automobile industry took notice of the topic as IEC 61508 got published as DIN EN 61508 (VDE 0803) "Functional safety-related electric/electronic/ programmable electronic systems" in 2001. Official correspondence between the VDA and the VDTÜVs led to the foundation of AK16 in FAKRA (Facharbeitskreis Automobil—German expert group from vehicle manufacturers and equipment suppliers), a group I became part of when I joined Continental Teves in 2004. In the same year, the first structures for the later ISO 26262 were designed and contact was established to further automobile standardization committees in other countries. Especially with France, concrete parameters for the standard were developed. The first meeting of the standardization group of ISO/TC22/SC03/WG16 took place from October 31 to November 2, 2015 in Berlin. The biggest delegate groups were from France and Germany besides representatives from other countries such as Japan, the USA, Sweden, Great Britain et cetera. Up to this point, ISO 26262 was still called 'FAKRA-Norm' (FAKRA-Standard). SafeTronic 2005 (Safety Event from Hanser-Verlag) already addressed the first ideas for future automobile standards and the presentations held included 'Best Practices' and methods. Until today, SafeTronic supported the development of ISO 26262, which got published as "International Standard" in November 2011. This book tries to compile all the background information that has been collected over the years. Moreover, it aims to give a better understanding of safety architecture as a basis for the development of safety-related products.

Preface

The following book is the result of over 20 years of professional experience in the field of functional safety. When I started my career after graduating as an engineer in 1992, plant engineering and construction was highly influenced by catastrophic events such as 'Bhopal' and 'Seveso'. The first set of rules and regulation which led later to IEC 61508 and ISO 26262 that addressed the issue of functional safety was the VDI/VDE guideline 2180 "Sicherung von Anlagen der Verfahrenstechnik; Safeguarding of industrial process plants by means of process control engineering" from 1966. However, it only covered the mere process of how to establish a safe environment in such facilities. In 1984 the differentiation between operational safety and safety equipment as well as monitoring and safeguarding equipment were added to the guideline. Thereafter, DIN VDE 31000—"General guide for designing of technical equipment to satisfy safety requirements" got published, which elaborated on the correlation between risk, safety and danger and introduced tolerable risk. At this time machinery standards, which prohibited the use of micro-controller for safety applications, were still common. However, an established market for safety-related control systems already existed. Different rules and standards defined the base of requirements for examinations, certifications and design of such systems. Those requirements were scaled in requirement classes (AK 1-8) according to DIN V 19250, independently from application or technology and explained a qualitative risk assessment procedure with the help of a risk graph.

In 1990 DIN V VDE 0801 "Principles for computers in safety-related systems" was released and in its revision of 1994 terms such as 'well-proven design principles' and the usage of 'consideration item' were added. By then, 'redundancy' was the only known answer to the various risk and requirement classes. However, various measuring principles were already used in measurement and control system engineering in order to detect hazardous situations early.

The technical rules for steam or the regulations for pressure vessels already required the redundant measurement of steam and temperature due to safety issues. Even the German Water Ecology Act mentioned the filling quantity limit from tanks according to regulations as well as the independent overfill safety device as a safety measure. A lot of those safety principals emerged from the safety standards of plant operators and even served as a foundation for official permits or releases. Even before in the early sixties DGAC (Direction General de L´Aviation Civil in France), CAA (Civil Aviation Authority) in Great Britain or FAA (Federal Aviation Administration) in USA and the military and space industry defined regulations about "Functional Safety", but those were not in the focus of the development of standards like IEC 61508 and ISO 26262. Due to today's discussion about 'autonomous' or 'automated' driving, those standards become more and more in the focus of the automotive industry. Especially topics such as safety-in-use, fail-operational, security, operational safety are becoming important for future revisions of ISO 26262.

In 1998, at the time I started my job as a sales manager of safety-related control systems, discussions over the early drafts of IEC 61508 took place, especially in countries such as England, the Netherlands and Norway. The scalable redundancy was a known concept so the discussion focused on the distinction between redundancy for safety and availability. Micro-controllers were coupled according to the lockstep principle and could change the program sequence or control logistics during runtime of a plant. Programming software was available, which allowed configuring the safety logic within a defined runtime environment.

The publication of IEC 61508 introduced a lifecycle approach for safety systems. Additionally, it formulated a process approach for product development and the relations to quality management systems were formulated.

During my graduate studies at the Faculty of Business and Economy at the University of Basel, I was able to hear a lecture of Prof. Dr. Walter Masing, who had a huge impact on quality management systems in Germany. The introduction of implemented diagnostics for the safety of functions and the electric carrier systems of these functions, respectively, broadened the view of safety architecture. In 1998, I introduced the first passive electronic system in Birmingham, which until SIL 4 was certified according to IEC 61508. I witnessed when the first certificate for a single-channel control system got signed after SafeTronic in 1999, which took place in the facilities of TÜV-Süd. This system was completely developed according to IEC 61508.

During VDMA-events (Verein Deutscher Machinen und Anlagenbauer; German machinery and plant engineering association) I reported on my experiences with IEC 61508 regarding plant engineering and its influence on the development of safety-related control systems. In these days, the machinery engineering industry was still heavily influenced by relay technology. Nobody wanted to believe that software-based safety technology would change the industry so drastically and in such a short time by providing new solutions and change existing systems. In 2001 I became the head of product management; the main task was to find new applications for new safety systems. Another main topic was 'safe network technology', which was so far based on serial link data busses. The challenge was to realize distributed and decentralized safety systems based on dynamic, or situation-, or condition-dependent safety algorithm. The only possible solution turned out to be 'Ethernet'. It was important to make the existing computer or data technology for safety technology easily manageable. In Norway, in the context of diploma theses, safety control systems got distributed, which exchanged safety-relevant data within the data network of the Norwegian mineral oil association “Statoil”. The experiences with the data transfer over satellites between oil platforms and plants ashore or between Norway and Germany as well as various solutions to the pipeline monitoring via radio systems proved that the safety technical data systems were also able to be realized based on Ethernet.

Hans-Leo Ross

Acknowledgments

The plentiful discussions with experts of international standardizations, colleagues, within the working groups, universities and presentations as well as the insights of diploma theses and public funding projects have contributed to this book. I would like to thank all the people involved for their shared passion for functional safety. Besides all the experts I especially want to thank my wife, who showed a lot of understanding and gave me the freedom and space to write this book.

Contents

Chapter 1
Introduction

ISO 26262 [1] changes vehicle development in a way, nobody would have expected 10 years ago, when functional safety became a relevant topic in the automobile industry. During the early 21st century the first German (VDA) working group already started dealing with functional safety and when the first international working groups got founded in 2005 everybody was looking for a lean standard for product safety. In the following 10 years before the final publication of the ISO 26262, those working groups compiled 10 parts with about 1000 requirements. Even though a lot of pertinent knowledge, methodologies and approaches have been discussed throughout the years, only a fracture of it has been incorporated in ISO 26262. Some information has only been added as footnotes, some disappeared completely until the final release of the standard.

In order to translate ISO 26262 there are currently various standardization projects in progress in different countries worldwide. The aim is to translate ISO 26262, provide further guidelines and develop additional methodologies for functional safety based on ISO 26262.

ISO 26262 is not intended to serve as a guideline it simply provides requirements for activities and methods, which should be taken into account in the respective functional safety activities. There is no description included as to how the requirements are supposed to be met. The underlying assumption is that such a state-of-the-art safety standard is considered to be a current up-to-date knowhow and will only be valid within a certain period of time. Recommendations on which designs are considered to be safe or which methodologies are adequate for certain activities are only valid and satisfactory until new or better methods are found. Also, safety design and methodology should be continuously improved and never limited to safety standards. There is an enormous need for guidelines and this book aims to provide further insights and background information on the respective topic but it does not offer guidelines on the correct application of ISO 26262. It focuses on methods and methodologies but none of those mentioned could fulfill the requirements of ISO 26262. Standards can only be fulfilled in the context of developing a real product in a given environment.

H.-L. Ross, *Functional Safety for Road Vehicles*,
DOI 10.1007/978-3-319-33361-8_1

Requirements, hints and notes in ISO 26262 are often described in a very complex way. The choice of words is a compromise experts who developed those safety standards had to agree upon. This is why all translations in this book may already be seen as interpretations, which could be interpreted or translated in other ways in the light of a different context. The strong recommendation to all readers is to reference to the text of ISO 26262 when trying to interpret and apply those standards in the field.

1.1 Definitions and Translations from the ISO 26262

ISO 26262 was only written in English. Even the usually common translation to French was not implemented due to the different use and interpretation of certain terms. This is why ISO 26262 is one of the only standards for which the original English text is also used in France. Asian countries are the only countries that have published a translation in their native language, a necessary requirement considering that the average developer in Japan would face difficulties in reading, understanding and interpreting the English language. After Japanese, Korean and Chinese translations followed afterwards during last years. For example, there is only one word in Japanese for verification, analysis, investigation and validation, thus the English text could have caused too many interpretation issues. Japanese translators assured that the content would not be falsified.

Finding the right and accurate words proved to be difficult even for the translation to the German language. Terms such as verification, analysis and validation were used in accordance with ISO 26262. However, some terms from the ISO 26262 glossary, highlighted in the blue boxes found throughout the book are citations from ISO 26262, but all explanations before or after are interpretations from the understanding of the author. Free interpretations, opinions or even recommendations of the author are written in the standard font chosen throughout the book; direct quotes are written in italics.

Throughout this book, the terminology "assessment of functional safety" is used to refer to the activity involved in "Functional Safety Assessment" as described in ISO 26262. In considering this concept of "assessment," it should be noted that "examination" is the basis for assessment and results in "judgment" of a property of the vehicle system or element.

ISO2626, Part 1, Clause 1.4:

1.4 (Assessment)
Examination of a property of a vehicle system (1.69) or element (1.32)
Note: A certain degree of independence (1.61) of a certain party or parties who perform an assessment should be ensured for each assessment.

The English word ‘assessment’ is translated as the German word used for ‘judgment’ and examination is seen as the basis for an assessment. The term “Assessment of Functional Safety” is used regarding the activity “Functional Safety Assessment” described in ISO 26262.

ISO2626, Part 1, Clause 1.6:

> 1.6 ASIL (Automotive Safety Integrity Level)
> One of four levels to specify the item’s (1.69) or element’s (1.32) necessary requirements of ISO 26262 and safety measures (1.110) to apply for avoiding an unreasonable residual risk (1.97), with D representing the most stringent and A the least stringent level.

For ‘Automotive Safety Integrity Level’, this book only uses the abbreviation ASIL.

ISO 26262 already provides a description of the elements of a vehicle system in part 10. An ‘element’ could be a system, a subsystem (logical or technical element and thus also a functional group), a component, a hardware device or a SW unit.

Part 1 of ISO 26262 is described under 1.69 Vehicle System (item) as follows:

ISO2626, Part 1, Clause 1.69:

> 1.69 (vehicle system, item)
> system (1.129) or array of systems to implement a function at the vehicle level, to which ISO 26262 is applied.

The word ‘item’ in English is often considered as ‘vehicle system’ in the context of this book, if it refers to the concrete word ‘item’ as used in ISO 26262 the word “ITEM” is used.

Historically, the German term for “Betrachtungseinheit” could be translated as “unit or item under consideration”. The English word ‘item’ and its definition as a system better applies to the idea of a vehicle system. Whenever this is relevant in the text, the term ‘item’ is added in brackets. The term ‘array of systems’ will be questioned in Chap. 4 of this book. A systematic hierarchical structured systems and associated subsystems are required in the technical parts from ISO 26262.

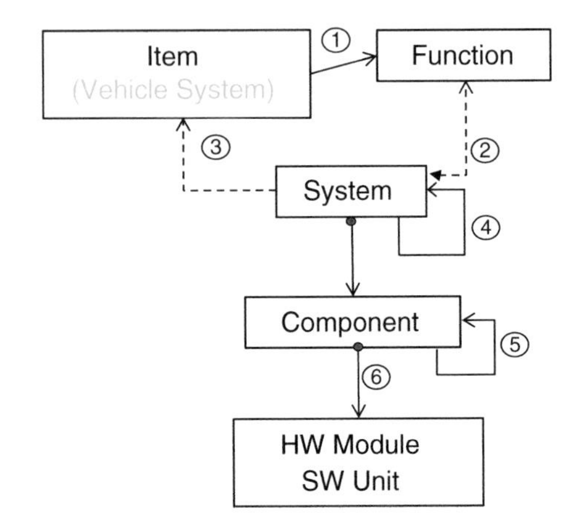

Fig. 1.1 Elements of a vehicle system (*Source* ISO 26262, Part 10, Fig. 3)

Figure 3 (here Fig. 1.1) from part 10 can be described as follows according to this definition:

ISO2626, Part 10, Fig. 3:

1. A system (1.129) or more systems, which realize one (or more) function(s) on the vehicle level for which ISO 26262 can be used.
2. A system may implement one or more functions, but also one function can be implemented in several systems.
3. A vehicle system is comprised of one or more systems, where one system is composed of at least one sensor, a processing unit and an actuator. ISO 26262 draws the conclusion that a system should have at least three elements but it could be possible for example that an actuator is integrated in the processing unit.
4. A system can be divided into any subsystems but according to ISO 26262 the systems have to be hierarchically structured. In regards to systems, which together should realize functions with a higher ASIL, a clear hierarchical structure of systems has to be defined due to multiple fault control.
5. A system (or subsystem) is comprised of one or more components.
6. Components consist of (electrical) hardware components (hardware parts) or of SW units.

Terms such as module, SW-files et cetera are not defined in ISO 26262. In regards to embedded semiconductors the term ‘Sub-Parts’ is used. Sub-Parts are logical functional elements, which implement specific functions and safety mechanisms within an integrated semiconductor.

1.2 Error Terms of the ISO 26262

ISO 26262 specifies terms in Volume 1 as follows:
ISO2626, Part 1, Clause 1.36, 39, 42:

> 1:36 (error)
> discrepancy between a computed, observed or measured value or condition, and the true, specified, or theoretically correct value or condition
> NOTE 1 An error can arise as a result of unforeseen operating conditions or due to a fault (1.42) within the system (1.129), subsystem or component (1.15) being considered.
> NOTE 2 A fault can manifest itself as an error within the considered element (1.32) and the error can cause a failure (1.39) ultimately.
>
> 1:39 (failure)
> termination of the ability of an element (1.32), to perform a function as required
> NOTE Incorrect specification is a source of failure.
>
> 1:42 (fault)
> abnormal condition that can cause an element (1.32) or an item (1.69) to fail
> NOTE 1 Permanent, intermittent and transient faults (1.134) (especially soft-errors) are considered.
> NOTE 2 An intermittent fault occurs time and time again—and disappears. These faults can happen when a component (1.15) is on the verge of breaking down or, for example, due to a glitch in a switch. Some systematic faults (1.131) (e.g. timing marginalities) could lead to intermittent faults.

The following assumptions were made due to different usages of the terms "Fault", "failure" and "error" in their context:

- Fault: Deviation, anomaly, defect, defect, non-conformity
- Error: mistake, fault or error
- Failure: Failure or malfunction.

The relationships of these three terms and also their model of error propagation are described in Sect. 4.4.2. Here only needs to be noted that the term "error" in German more generally and is thus used in this book primarily as a collective term for all three terms. If error purely regarded as "error", this is explained in the context.

In the safety analysis the following aspects can be distinguished:

- Single point fault (or single failure) and
- Multiple-point faults.

If a single fault or a deviation of an observable behavior or property alone leads to a failure of a system, this is referred to as a single point fault. Perform only a combination of several faults, deviations to unintentional changes, observable behavior or changed properties; this is regarded as a multiple-point faults. At least combinations of minimum two faults are necessary to propagate to a multiple-point failure. In ISO 26262 this naming is not based on a system's behavior, but on a safety goal. For example single point faults are considered only if a single fault leads to a violation of the safety goal within the specified Item, boundary or the specified environment or specified space. Faults which lead only to failure outside the "Item" are not considered as a single point fault, unless the "Item Definition" not changed due to systematic failures.

References

1. [ISO 26262]. ISO 26262 (2011): Road vehicles – Functional safety. International Organization for Standardization, Geneva, Switzerland.

ISO2626, Part 1, Clause 1.4: 2

ISO2626, Part 1, Clause 1.6: 3

ISO2626, Part 1, Clause 1.69: 3

ISO2626, Part 10, Figure 3: 4

ISO2626, Part 1, Clause 1.36, 39, 42 5

Chapter 2
Why Functional Safety in Road Vehicles?

It took a while until functional safety started to play a significant role in the automotive industry in comparison to other industries. Customers, producers and dealers networks demanded more functionality and complexity of the products and market. One of the major reasons was that mechanical engineers primarily dominated the entire automobile engineering industry. The same industry developed the safety mechanism in the related field, without relying on electronics or even software. Therefore, these safety mechanisms were first and foremost based on a robust design as well as hydraulic or pneumatic safety mechanisms. With the increased amount of automation and electrification of essential vehicle functions and the desire to make these systems applicable for higher speeds and dynamics, electrification was the only way to go. Also the earlier concepts steer-by-wire and brake-by-wire, right up until today's autonomous or highly automated driving systems, make the usage of software based safety mechanisms unavoidable. If you look at one of today's common mid-range cars such as the 'Volkswagen Golf', you will find about 40 control units, which are still mainly networked by a CAN-Bus. It is "State-of-Science and Technology" that no complex vehicle systems could be realized without a systems approach. One of the main challenges of ISO 26262 [1] was that various methods, methodology, principles, best practices had been established but there was no consistent system development approach.

The main task in the development of ISO 26262 was to agree upon one basic understanding of system engineering. Therefore, it is not a surprise that the word 'system engineering' appears quite often in the introduction.

2.1 Risk, Safety and Functional Safety in Automobiles

In general, risk is described as a possible event with a negative impact. The Greek origin of the word risk had been also used for hazard or danger. In regards to product safety it is referred to as the cross product of probability of occurrence and

H.-L. Ross, *Functional Safety for Road Vehicles*,
DOI 10.1007/978-3-319-33361-8_2

hazard/danger. There are different opinions on the term and definition of risk in the economic literature. Definitions vary from 'danger of a variance of error' to the mathematical definition 'risk = probability × severity'.

The general definition is as follows: The probability of damage or loss as consequence of a distinct behavior or events; this refers to hazardous/dangerous situations in which unfavorable consequences may occur but do not necessarily have to.

On the one hand, risk can be traced back etymologically to 'riza' (Greek = root, basis); see also 'risc' (Arabic = destiny). On the other hand, risk can be referred to 'ris(i)co' (Italian); "The cliff, which has to be circumnavigated". 'Safety' derives from Latin and could be translated as 'free from worry' (se cura = without worry). Today, the topic of safety is viewed in various different contexts for example, in regards to economic safety, environmental safety, admittance and access security but also in terms of work safety, plant and machinery safety and vehicle safety. The term safety varies significantly from just only functional safety.

In relation to technical systems or products, safety is described as the freedom of unacceptable risks. 'Damage' is generally seen as harm or impairment of people as well as the environment.

There are various distinctions of hazard:

- Chemical reactions of substances, materials etc. lead to fire, explosions, injuries, health impairments, poisoning, environmental damage etc.
- Toxic substances lead to poisoning (also carbon monoxide), injuries (consequence of for example degassing of batteries, error reactions of the driver or mistakes of the auto repair shop staff), other damages etc.
- High currents and especially high voltages lead to damages (in particular personal protection).
- Radiations (nuclear, but also radiations like alpha particle semiconductor).
- Thermic (damages due to overheating, singe, fire, smoke etc.).
- Kinetics (deformation, movement, accelerated mass can lead to injuries).

The potential reasons for hazard cannot be easily defined, since chemical reactions can also lead to poisoning and overheating, to fire and thus also to smoke intoxication. Similar correlations appear in high currents or excessive voltages. High voltages lead to burns when touching but can also cause fires. Overvoltage is often seen as a non-functional risk or hazard. This is why most of the standards encounter such hazards with design constraints. A contact safety device or touch guard on a safety plug connector is a typical example. This leads us to the following point of view and distinction of functional safety.

Functional safety is generally described as the correct technical reaction of a technical system in a defined environment, with a given defined stimulation as an input of the technical system. ISO 26262 defines functional safety as absence of unreasonable risk due to hazards caused by malfunctioning behavior of E/E systems. Also, the error or failure reactions of mechanic or hydraulic safety components are

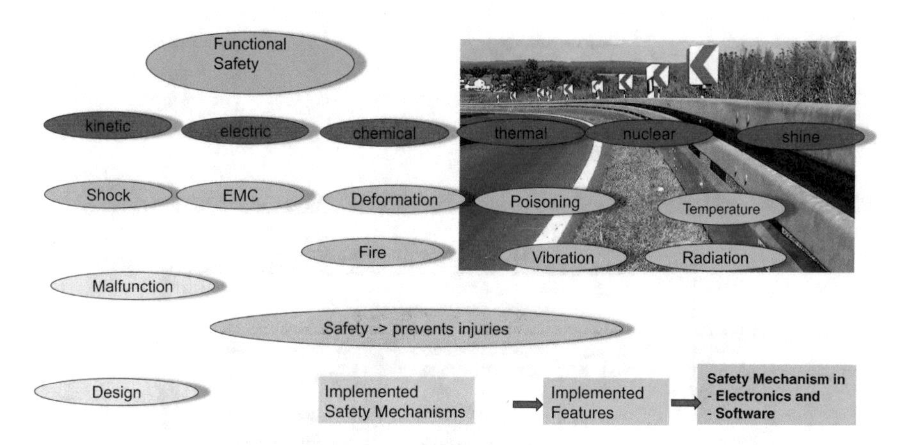

Fig. 2.1 Functional safety—safety design, control of forces and energies

controlled by electronic safety mechanisms in mechatronic systems. This distinction will be discussed later in reference to ISO 26262 (Fig. 2.1).

Functional safeguarding with hydraulic systems has always been used for automobiles. A typical example would be the dual-circuit braking system or the hydraulic steering system. Electronic and software based functional safety mechanisms were introduced as for example the ABS (Anti-Wheel-Blocking-System) for brake systems 30 years ago. Prior to that the necessary safety was only established by sufficient robust system and safe component characteristics (meaning through design).

The following definitions of risk, hazard/danger and integrity have been added to *DIN EN 61508-1:2002–11:*

Citation from IEC 61508 [2], *Part 5, A5:*
A.5 Risk and Safety Integrity

It is important that the distinction between risk and safety integrity be fully appreciated. Risk is a measure of the probability and consequence of a specified hazardous event occurring. This can be evaluated for different situations [EUC risk, risk required to meet the tolerable risk, actual risk (see Fig. A.1)]. The tolerable risk is determined on a societal basis and involves consideration of societal and political factors. Safety integrity applies solely to the E/E/PE safety-related systems, other technology safety related-systems and external risk reduction facilities and is a measure of the likelihood of those systems/facilities satisfactorily achieving the necessary risk reduction in respect of the specified safety functions. Once the tolerable risk has been set, and the necessary risk reduction estimated, the safety integrity requirements for the safety-related systems can be allocated. (see 7.4, 7.5 and 7.6 of IEC 61508-1) (Fig. 2.2).

Furthermore, IEC 61508 shows the following figure to explain coherences (Fig. 2.3):

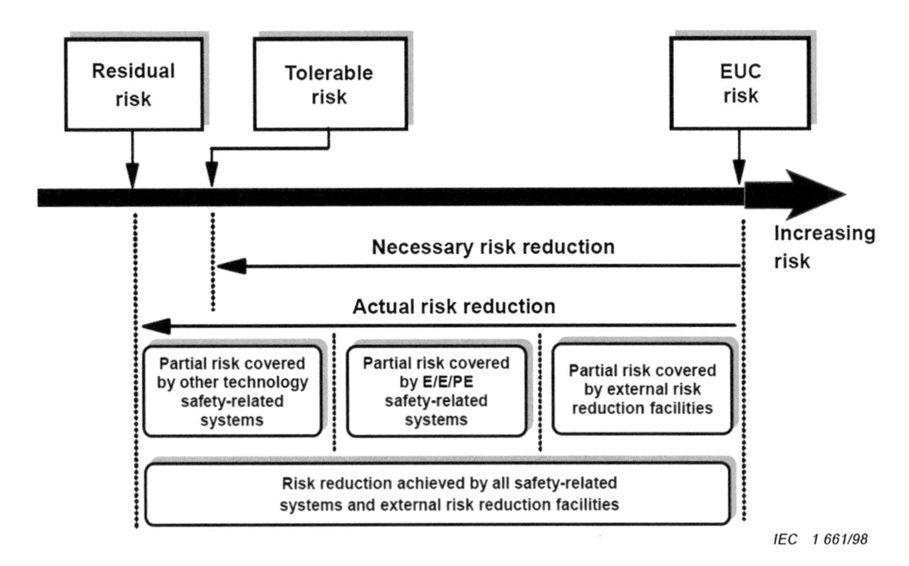

Fig. 2.2 Risk reduction according to IEC 61508 (*Source* IEC 61508-1:2011)

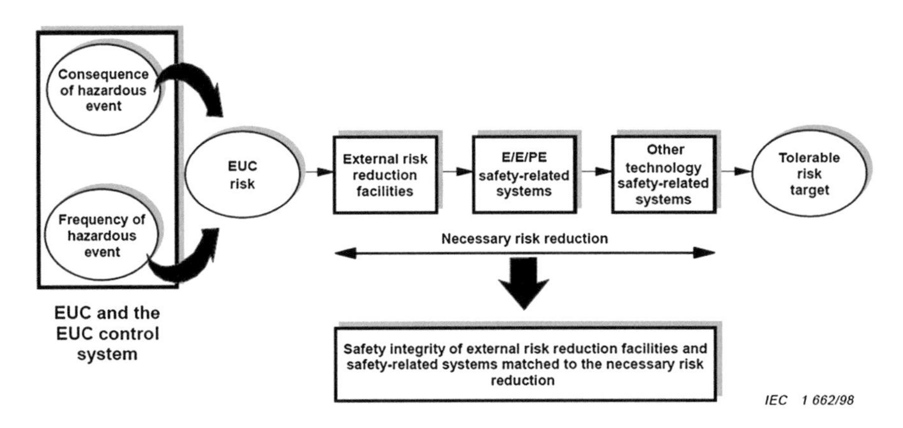

Fig. 2.3 Risk- and safety integrity according to IED 61508 (*Source* IEC 61508-1:2011)

ISO 26262 defines the relation of risk, danger and safety integrity differently. The term safety integrity is not directly used in ISO 26262. In particular the term EUC (Equipment under Control) is not used at all. EUC could be explained as "device or system, which should be controlled by means of functional safety measures". Under certain limiting conditions ISO 26262 admits to develop a desired vehicle function that is safety-related on its own. In this case, the system does not receive safety through EUC itself. Technically, according to IEC 61508, EUC and the safety functions have to cause an error at the same time in order to create a hazardous situation. If for example a hydraulic braking system was the

EUC, which in its function can be monitored by an EE-system, errors of the hydraulic systems could be avoided by the EE-system. The automobile industry relies here on other technology and engineering of the electronic safety system will be considered as a fail-safe-system.

As mentioned previously, ISO 26262 defines functional safety as freedom of unacceptable risks based on hazards, which are caused by malfunctional behavior of E/E-systems. However, interactions of systems with E/E-functions are included as well and therefore also mechatronic systems. Whether pure mechanical systems really show not any interactions with E/E is doubtful. Furthermore, the introduction chapter of ISO 26262, which describes the scope of the norm, excludes hazards such as electric shock, fire, smoke, heat, radiation, poisoning, inflammation, (chemical) reactions, corrosion, release of energy or comparable hazards, as long as the failure was not caused by electrical components. Such hazards are caused more by the battery as well as the poisonous electrolytes in the capacitors. Whether a motor winding is an electrical device or a mechanical component is also questionable.

In general, it will be difficult to assign the ASIL with non-functional hazards. Such components have so far been construed sturdily in order to avoid any danger. In the context of the hazard and risk analysis it is difficult to allocate a specific ASIL to a weakness in design or construction.

ISO 26262 also excludes functional performances. Therefore, safety-in-use or functional inadequacy means functions, which already lead to a hazard, even if they functioning correctly are generally excluded in advance.

All explain the correlation of risk and damage as follows:

ISO 26262, part 3, appendix B1:

For this analytical approach a risk (R) can be described as a function (F), with the frequency of occurrence (f) of a hazardous event, the ability of the avoidance of specific harm or damage through timely reactions of the persons involved (controllability: C), and the potential severity (S) of the resulting harm or damage:

$$R = F(f, C, S)$$

The frequency of occurrence f is, in turn, influenced by several factors. One factor to consider is how frequently and for how long individuals find themselves in a situation where the aforementioned hazardous event can occur. In ISO 26262 this is simplified to be a measure of the probability of the driving scenario taking place in which the hazardous event can occur (Exposure: E). Another factor is the failure rate of the item that could lead to the hazardous event (Failure rate: λ). The failure rate is characterized by hazardous hardware random failures and systematic faults that remained in the system:

$$f = E \times C$$

Hazard analysis and risk assessment is concerned with setting requirements for the item such that unreasonable risk is avoided.

ISO 26262 mentions normative methods that describe a systematic derivation of the potential risk, which may originate from the investigated of the considered Item (vehicle system), based on a hazard analysis and risk assessment. Hazard or risk analyses are not normatively defined in other safety standards. Either the requirements for these methods are listed or the method itself is exemplarily described (Fig. 2.4).

The reduction of risk cannot be achieved with the activities and methods mentioned in ISO 26262 if a function is not suitable, inadequate suitable, inadequate or falsely indicated for certain safety related functions. This represents a special challenge, considering that ISO 26262 does not directly addresses a EUC (Equipment under Control, e.g. a system, machinery or vehicle, which should be controlled safety-related systems) or the distinction between safety functions of designated safety requirements for on-demand (low demand) or continuous mode (high demand) safety systems. How is it possible to find out whether or not reactions of a vehicle system or certain measurements are sufficient, tolerable or safety-related appropriate?

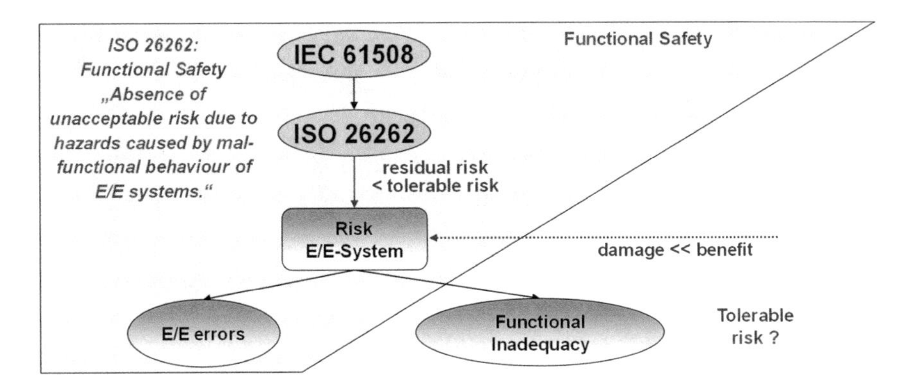

Fig. 2.4 Distinction of hazards, based on correctly functioning systems (*Reference* unpublished research project [7])

2.2 Quality Management System

Prof. Dr. rer. nat. Dr. oec. h. c. Dr.-Ing. E. h. Walter Masing, is also called the father of quality management systems, at least in Germany. His standard reference "Masing Handbook Quality Management" had a substantial influence on the standardization and interpretation of quality management systems.

A lot of methods and principles of management systems are explained already in ISO 9000. However, in 2005, statistics and trial methods became less relevant as the process approach became more and more important.

In the automotive industry an addition to ISO 9001 exists, called ISO TS 16949 [3]. It describes additions especially to the product development and production, which developed into standards in this industry. Today, in order for a distributor to be able to supply automotive manufacturers, the certification of ISO TS 16949 is an essential basic. Manufacturers from Asia still refer to different standards, based on historical reasons. Especially in Japan, quality requirements focus more on the ideals of the six-sigma-philosophy (for example DFSS, Design for Six Sigma). In particular the static analysis and trial methods mentioned in Masing's book, in DSFF as well as in functional safety are often based on comparable principles. ISO TS 16949 asks in the following chapters for essential basics for functional safety according to ISO 26262:

ISO TS 16949, 4.2.3.1: Engineering specifications

The organization shall have a process to assure the timely review, distribution and implementation of all customer engineering standards/specifications and changes based on customer-required schedule. Timely review should be as soon as possible, and shall not exceed two working weeks.

The organization shall maintain a record of the date on which each change is implemented in production. Implementation shall include updated documents.

NOTE A change in these standards/specifications requires an updated record of customer production part approval when these specifications are referenced on the design record or if they affect documents of production part approval process, such as control plan, FMEAs, etc.

Here, the norm refers to document and change management, application of necessary norms and standards, methods, output/work results and the regulation of responsibility (clearance), which is mentioned in ISO 26262 as QM-methods.

ISO TS 16949, 5.6.1.1 Quality management system performance

These reviews shall include all requirements of the quality management system and its performance trends as an essential part of the continual improvement process.

Part of the management review shall be the monitoring of quality objectives, and the regular reporting and evaluation of the cost of poor quality (see 8.4.1 and 8.5.1).

These results shall be recorded to provide, as a minimum, evidence of the achievement of

- *the quality objectives specified in the business plan, and*
- *customer satisfaction with product supplied.*

This explains the fact that product development as well as the satisfaction of the products delivered has to be documented and proven. If it concerns safety related features this may affect the customer substantially.

ISO TS 16949, 5.6.2: Review input
ISO 9001:2000, Quality management systems—Requirements

5.6.2 Review input
The input to management review shall include information on

a) results of audits,
b) customer feedback,
c) process performance and product conformity,
d) status of preventive and corrective actions,
e) follow-up actions from previous management reviews,
f) changes that could affect the quality management system, and
g) recommendations for improvement.

This list can also be seen as a “safety culture” in infrastructure requirements and essential for functional safety.

ISO TS 16949, 5.6.2.1: Review input

Input to management review shall include an analysis of actual and potential field-failures and their impact on quality, safety or the environment.

This chapter refers directly to the essential field observations, which are also required by the government in the context of product liability laws. It also directly refers to safety defects.

ISO TS 16949, 5.6.3: Review output
ISO 9001:2000, Quality management systems—Requirements

5.6.3 Review output
The output from the management review shall include any decisions and actions related to

a) improvement of the effectiveness of the quality management system and its processes,
b) improvement of product related to customer requirements, and
c) resource needs.

There are further additions mentioned to this topic in particular in ISO 26262.

ISO TS 16949, 6: Resource management

6.1 Provision of resources

ISO 9001:2000, Quality management systems—Requirements 6 Resource management 6.1 Provision of resources The organization shall determine and provide the resources needed (a) to implement and maintain the quality management system and continually improve its effectiveness, and (b) to enhance customer satisfaction by meeting customer requirements.

6.2 Human resources

6.2.1 General

ISO 9001:2000, Quality management systems—Requirements 6.2 Human resources 6.2.1 General

Personnel performing work affecting product quality shall be competent on the basis of appropriate education, training, skills and experience.

Sections 6.1 and 6.2 show, that also in the development stage essential requirements of people, their qualifications and the organization of product creation are well defined according to quality management systems.

ISO TS 16949, 7.3.1.1: Multidisciplinary approach

The organization shall use a multidisciplinary approach to prepare for product realization, including

- *development/finalization and monitoring of special characteristics,*
- *development and review of FMEAs, including actions to reduce potential risks, and*
- *development and review of control plans.*

NOTE A multidisciplinary approach typically includes the organization's design, manufacturing, engineering, quality, production and other appropriate personnel.

This cross-functional approach of ISO TS 16949 defines the basis for a necessary safety culture as the foundation of functional safety and address directly FMEAs as a mayor quality analysis method.

ISO TS 16949, 7.3.2.3: Special characteristics

The organization shall identify special characteristics [see 7.3.3 d] and

- *include all special characteristics in the control plan,*
- *comply with customer-specified definitions and symbols, and*
- *identify process control documents including drawings, FMEAs, control plans, and operator instructions with the customer's special characteristic symbol or the organization's equivalent symbol or notation to include those process steps that affect special characteristics.*

NOTE Special characteristics can include product characteristics and process parameters.

This chapter defines the way safety requirements were handled previously in the automobile industry. In particular "special characteristics" are still used for a safety-related design parameter of mechanic parts. The paragraph also defines the basics for the production of safety related components.

ISO TS 16949, 7.3.3.1: Product design output—Supplemental

The product design output shall be expressed in terms that can be verified and validated against product design input requirements. The product design output shall include

- *Design FMEA, reliability results,*
- *product special characteristics and specifications,*
- *product error-proofing, as appropriate,*
- *product definition including drawings or mathematically based data,*
- *product design reviews results, and*
- *diagnostic guidelines where applicable.*

This is a list of the output of product development, which had to be extended in ISO 26262 for the relevant safety related work-products and components. This output would for example be part of the safety case in a safety related product development.

ISO TS 16949, 7.3.3.2: Manufacturing process design output

The manufacturing process design output shall be expressed in terms that can be verified against manufacturing process design input requirements and validated. The manufacturing process design output shall include

- *specifications and drawings,*
- *manufacturing process flow chart/layout,*
- *manufacturing process FMEAs,*
- *control plan (see 7.5.1.1),*
- *work instructions,*
- *process approval acceptance criteria,*
- *data for quality, reliability, maintainability and measurability,*
- *results of error-proofing activities, as appropriate, and*
- *methods of rapid detection and feedback of product/manufacturing process nonconformities.*

This list adds to the necessary output/work-products during production. ISO 26262 rarely mentions any further requirements since this area is well regulated by quality management systems.

ISO TS 16949, 7.5.1.1: Control plan

The organization shall

- *develop control plans (see annex A) at the system, subsystem, component and/or material level for the product supplied, including those for processes producing bulk materials as well as parts, and*
- *have a control plan for pre-launch and production that takes into account the design FMEA and manufacturing process FMEA outputs. The control plan shall*
- *list the controls used for the manufacturing process control,*
- *include methods for monitoring of control exercised over special characteristics (see 7.3.2.3) defined by both the customer and the organization,*
- *include the customer-required information, if any, and*
- *initiate the specified reaction plan (see 8.2.3.1) when the process becomes unstable or not statistically capable. Control plans shall be reviewed and updated when any change occurs affecting product, manufacturing process, measurement, logistics, supply sources or FMEA (see 7.1.4).*

NOTE Customer approval may be required after review or update of the control plan.

ISO TS16949 describes the requirements of production control regarding the precedent development and required analyses, for example FMEAs, in detail. Analyses for product development are required—even if these products can be developed according to quality management systems but without any safety requirements.

2.2.1 Quality Management Systems from the Viewpoint of ISO 26262

Quality management is not mentioned very consistent in ISO 26262. The requirements set are the fundamentals, which enable any functional safety method to be applied in the automotive industry. Content wise, the appendix of part 2 ISO 26262 raises many interesting topics covering safety culture. ISO 26262 shortly summarizes the fundamental requirements as follows:

ISO 26262 Part 2, Clause 5.3.2:

5.3.2 Further supporting information
5.3.2.1 The following information can be considered:

- *existing evidence of a quality management system complying with a quality standard, such as ISO/TS 16949, ISO 9001, or equivalent.*

ISO 26262 Part 2, Clause 5.4.4:

Quality Management during the Safety Lifecycle
The organizations involved in the execution of the safety lifecycle shall have an operational quality management system complying with a quality standard, such as ISO/TS 16949, ISO 9001, or equivalent.

This means, a well-organized, well-established and well-applied quality management system is the basis of functional safety. ISO TS 16949 is required of all vehicle manufacturers worldwide. Quality management systems without this can be disregarded. ISO 26262 also mentions several work-products where safety aspects, enhancements, or improvements need to be added to already consider work-products defined by the quality management system.

In further additions ISO TS 16949 provides the following definition for quality:

ISO TS 16949, addition:

Quality is defined as "the sum of characteristics of an entity regarding its suitability to fulfill defined and predetermined requirements". The term "entity" is here very vague. It is defined as follows: "Something that can be described and observed individually." Thus quality refers to characteristics and features of a finished product. In general it is assumed that these characteristics remain for a certain time after the production. Often this time period equals the warranty period. As long as it is stated in the specifications that the existing characteristics and features after the production should remain through the defined usage period, reliability is a part of quality.

This definition clearly asks for the concept of lifecycles as a requirement, hence quality features such as safety should be self-evident.

2.3 Advanced Quality Planning

ISO TS 16949 can be interpreted differently for the individual cases of application. This is why automobile manufacturers have since defined standards in order to guarantee quality in product development. Later, the American manufacturers such as Ford, GM or Chrysler met at AIAG to define joint requirements for quality management. In Germany, similar standards and requirements were developed under the umbrella of VDA. The aim was to define processes for the development as well as planned advanced product quality improvement measures, APQP (advanced product quality planning according to AIAG).

VDA and AIAG published a series of documents, which are considered to be the foundation for VDA- or AIAG members. Those various volumes of these documents are often mandatorily referenced in the contract documents for supplier.

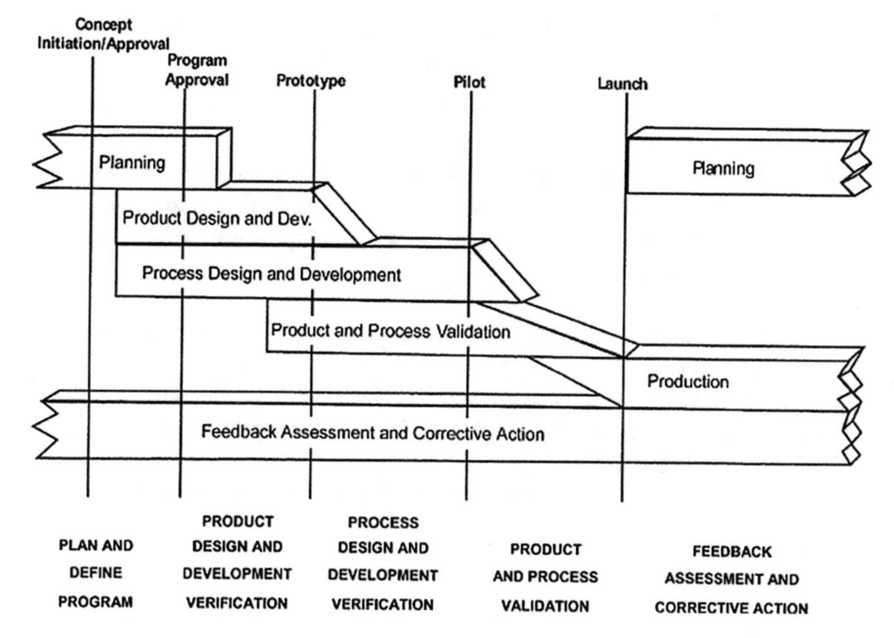

Fig. 2.5 Advanced product quality planning (*Reference* APQP AIAG [9] 4th Edition)

Unfortunately, these documents are not highly consistent. For example, both organizations describe different FMEA methods (or several FMEA methods), which are considered to be a basis of ISO 26262. In addition, these organizations also developed milestones or maturity level concepts, which were primarily used for the synchronization of automotive manufacturers and supplier (Fig. 2.5).

AIAG defined APQP with 5 "milestones":

- The first phase "Concept, initiation, approval" is a mere planning phase
- In the second phase, before the program approval, the planning as well as the product and process development should have a certain maturity. The feasibility of the product is then verified as part of the program approval.
- The third phase focuses on the development of the first prototypes, the verification (often prototype tests) and the product and production process validation. At this point, the product design should be almost finalized.
- In the fourth phase the first series-development (close-to-series-production, pilot) products are produced. Those products should already be produced with the series-production tools.
- The product launch initiates the series production. This requires the development of supply chains and the production needs to be able to guarantee a sufficient quantity and quality.

After the product launch an assessment of the product development and appropriate corrective actions are expected. All activities are continuously monitored and necessary corrective actions need to be implemented when field findings arise.

Within this topic VDA published the following volumes:

- *VDA QMS Volume 1*
  Documentation and Archiving—Code of practice for the documentation and archiving of quality requirements and quality records/3rd edition 2008
  Guidelines for documentation and archiving of quality requirements and records (especially for critical features).
- *VDA QMS Volume 2*
  Quality Assurance for Supplies Production process and product approval PPA, 5th revised edition, November 2012
  Choice of suppliers, quality assurance and agreements, production process and product approval, choice of ingredients (A new edition will be published soon)
- *VDA QMS Volume 3, part 1*
  Ensuring reliability of car manufacturers—Reliability Management/3rd edition 2000
- *VDA QMS Volume 3 part 2*
  Ensuring reliability of car manufacturers and suppliers—Reliability Management Methods and Utilities/3rd edition 2000, currently 2004
- VDA QMS Volume 4 Chapter Product and Process FMEA [4]
  2nd edition December 2006, updated in June 2012, (The chapter is already included in Volume 4)

These volumes are continuously updated and include new topics such as maturity level of products and processes, standardized requirement specifications etc.

2.4 Process Models

Procedure or process models have a long history. The following list shows the origin of such products, especially software-intensive products.

1. First attempt to develop clearly understandable programs (1968)
   Dijkstra suggests "structured programming" (Avoidance of GOTO-instructions).
2. Development of software engineering principles (1968–1974)
   Theoretical basics (principles) are developed that represent the foundation of structured development of programs: structured programming, step-by-step refining, secrecy concepts, program modularization, software lifecycles, entity relationship model, and software ergonomics
3. Development of phase-specific software engineering methods (1972–1975)
   Implementation of software engineering concepts in draft methods: HIPO, Jackson, Constantine method, first version of small talk

4. Development of phase-specific tools (1975–1985):
   Application of SE-methods with mechanic support (e.g. Program inversion, batch tools)
5. Development of phase-comprehensive (integrated) software engineering methods (since 1980)
   The results of one phase of the software lifecycle should be automatically passed on to the next phase: integrated methods
6. Development of phase-comprehensive (integrated) tools (since 1980)
   Application of databases as automatic interfaces between the individual phases of a software lifecycle. Interactive program cue through CAS-tools (computer aided software design)
7. Definition of different, competing and object oriented methods (since 1990)
   Various object oriented analyses and design methods were developed simultaneously. (Booch, Jacobson, Rumbaugh, Shlaer/Mellor, Coad/Yourdon et al.) These methods were implemented with CASE tools (computer aided software engineering)
8. Integration of OO-methods for UML-unified modeling language (since 1995)
   Jacobson, Booch and Rumbaugh joint to develop UML. UML aims to eliminate the previous weaknesses of OO-methods and create an internationally valid and uniform standard. UML 1.0 passed 1997.
9. UML 2.0
   UML 2.0 was published in 2004 after UML 1.0 was upgraded to version 1.5. This version includes adapted up to date language elements for new technologies and removed redundancies and inconsistencies in language definitions.
   Source: Online list without sources

History shows, that these approaches are merely based on experience. Over time, restrictions in the programming process have led to formalized description formats. Later, the description of these "best practices" as formalized activities lead to the development of process models as reference models or, as the example of UML shows, formalized description language. Certain principles such as, requirements are only accepted if they can be implemented and if tests show that they can be implemented correctly, influenced this strategic approach.

2.4.1 V-Models

The following figure shows the development of process models and process improvement models such as CMM or SPICE. ISO 9001 and ISO 12207 can be seen as a basis for these models. ISO 12207 is mentioned in the bibliography of ISO 26262. However, the relation between ISO 12207 and ISO 26262 is not explained.

Surprisingly, for a long time the principles of the process approach for product development have not been strongly developed in Asia. ISO 12207 is the foundation of process assessment models (PAM) based on CMM or SPICE. The practice of

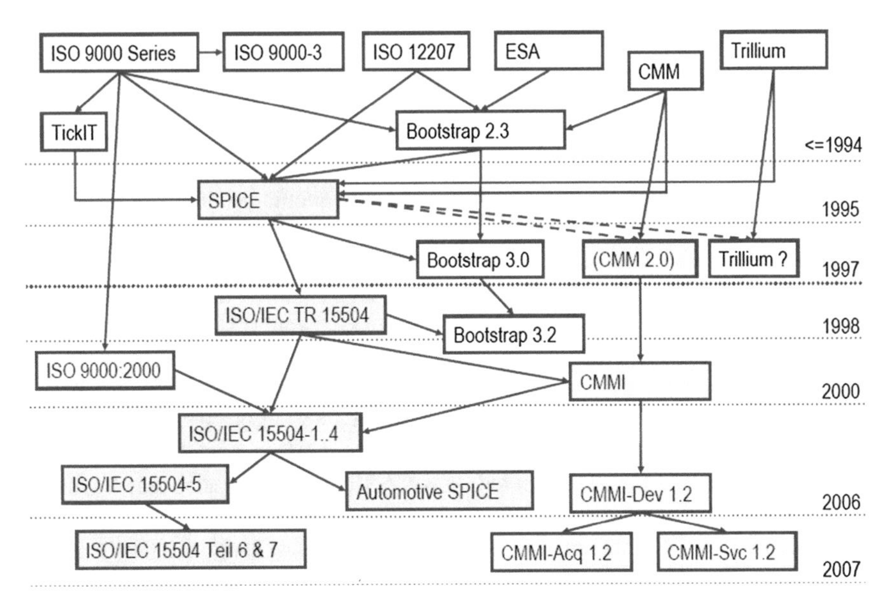

Fig. 2.6 History of procedure models based on V-models based (*Source* Flecsim)

relating those process assessment models with the safeguarding of software features was developed later.

The crucial question is 'does such a generic process actually represent more than what the SPICE-definitions describes?' Here the V-model is mentioned as a reference model. So if requirements of the development activities are described, is it useful to structure them according to such a reference model? (Fig. 2.6).

The V-model XT, in its version 1.2, describes the V only for the development of individual products (Fig. 2.7).

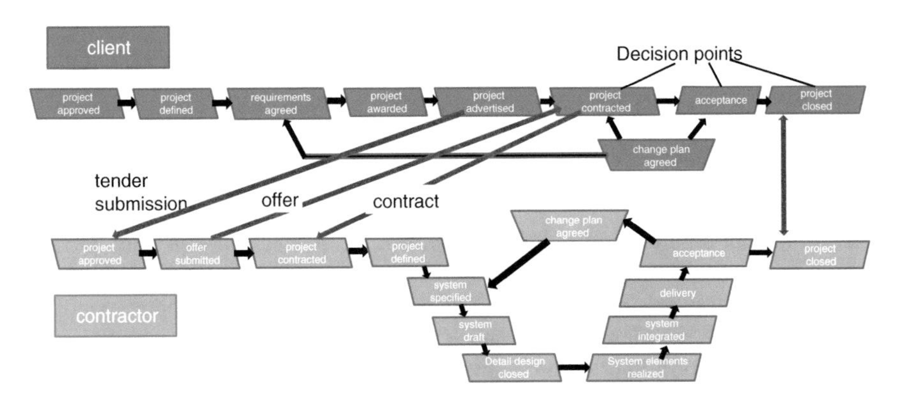

Fig. 2.7 Interface V-model customer—supplier according to V-model XT (*Source* V-model XT 1.2 [8])

The V-model XT first describes the customer—supplier relationship. This phase determines the product scope and the fundamental requirements and is comparable to part 8, Chap. 5 (Interfaces of the distributed development). Here the author refers to the interface agreement (DIA, Development Interface Agreement) between development partners. Those agreements should determine who is responsible for the various product development packages (or product elements) and who performs which activity (who does what).

SPICE (Software Process Improvement and Capability Determination) is often associated with ISO 26262 and is mainly based on two norms, ISO 12207 and ISO 15504.

ISO 12207 "Processes in software lifecycles" offers a process reference model with the following categories:

- customer-supplier processes,
- development processes,
- supportive processes,
- management processes,
- organization processes.

Part 6 of ISO 26262 mentions ISO 12207 in the bibliography appendix but there is no reference or explanation as to what relation those norms have with each other.

However, 40 processes are described, these are seen as a foundation for SW based product development and ISO 15504 derived a process assessment model (PAM) from this description.

ISO 15504 consists of the following parts:

ISO 15504-1: Concepts and vocabulary
Terms and general conception
ISO 15504-2: Implementation of assessments

- requirements for a process reference model
- requirements for PAM
- definitions of a framework to measure process capability levels
- requirements for an assessment process framework

ISO 15504-3: Guideline for the assessment implementation
Guideline for the implementation of a ISO 15504-2 conform assessment:

- Assessment framework for process capability levels
- PRM and PAM
- Selection and usage of assessment tools
- competence of assessors
- examination of compliance

ISO 15504-4: Guidelines for the usage of assessment results

- Selection of PRM
- Setting target capability

- definition of assessment inputs
- Steps to process improvement
- Steps to the determination of ability levels
- Comparability of assessment outputs

ISO 15504-5: Exemplary process assessment model (PAM)

Exemplary PAM, which fulfills all requirements of ISO 15504-2, and information on assessment indicators

ISO 15504-6: Exemplary PAM ISO 15228

- Structure of PAM
- Process performance indicators
- Process ability indicators

ISO 15504-7: Guidelines for the determination of the maturity level of an organization

CMMI and SPICE always differed in their assessments. SPICE always assesses individual processes but was unable to measure the maturity level of an organization like CMMI does. CMMI combines certain processes and therefor derives a maturity level for organizations.

With ISO 15504-7 also SPICE supports maturity levels for organizations.

ISO 15504-8: Exemplary process assessment model (PAM) for ISO 20000

Exemplary PAM for the IT service management

ISO 15504-9: Process profile goals

Part 9 is a technical specification (TS) which describes process profiles.

ISO 15504-10: Safety Extensions

Aspects of safety

AutoSIG used ISO 15504 as a basis for Automotive SPICE®. Part 2 and 5 were used for PAM and PRM. Automotive SPICE® is an adaption of parts of ISO 15504 to automotive applications.

Further lifecycle approaches for the SW development:

- ISO/IEC/IEEE 16326 Systems and software engineering—Lifecycle processes —Project management (2009)
- SAE J2640, General Automotive Embedded Software Design Requirements (April 2006)
- IEEE STD829, Standard for Software and System Test Documentation (2008)
- ISO/IEC 9126 Software engineering—Product quality (2001)
- ISO/IEC 15288 Systems engineering—System lifecycle processes (2002)
- ISO/IEC 26514 Systems and software engineering—Requirements for designers and developers of user documentation (2008)

All these norms influenced the development of ISO 26262. However, none of these norms from the list above is in a normative relationship with ISO 26262.

However, the norms of the ISO/IEC 25000 [5] were highly influential. They were developed simultaneously to ISO 26262 and since 2005 have replaced ISO/IEC 9126.

The basic norm is called:

ISO/IEC 25000 Software engineering—Software Product Quality Requirements and Evaluation (SQuaRE)

This series includes quality criteria and the ISO organization asks other norm developing working groups to use these as guidelines.

The following examples show a comparison of the definitions of ISO/IEC 25000 and ISO 26262:

Functionality:

The capability of the software product to provide functions, which meet stated and implied needs when the software is used under specified conditions.

Generally does not contradict with ISO 26262

Functional appropriateness:

Degree to which the functions facilitate the accomplishment of specified tasks and objectives. EXAMPLE An user is only presented with the necessary steps to complete a task, excluding any unnecessary steps.

NOTE Functional appropriateness corresponds to suitability for the task in ISO 9241-110.

The term is not used in ISO 26262, but does not mean any contradiction.

Functional correctness:

Degree to which a product or system provides the correct results with the needed degree of precision.

Considered as part of verification measures, but not addressed as such in ISO 26262.

Interoperability:

Degree to which two or more systems, products or components can exchange information and use *the information that has been exchanged*

NOTE Based on ISO/IEC/IEEE 24765.

The focus in ISO 26262 lies more on the flawed cooperation of elements and systems.

The term is not used in ISO 26262, but does not mean any contradiction.

Security:

Degree to which a product or system protects information and data so that persons or other products or systems have the degree of data access appropriate to their types and levels of authorization.

NOTE 1: As well as data stored in or by a product or system, security also applies to data in transmission.

NOTE 2: Survivability (the degree to which a product or system continues to fulfill its mission by providing essential services in a timely manner in spite of the presence of attacks) is covered by recoverability (4.2.5.4).

NOTE 3: Immunity (the degree to which a product or system is resistant to attack) is covered by integrity (4.2.6.2).
NOTE 4: Security contributes to trust (4.1.3.2).

The term is not yet addressed, but it is a big topic for future revisions of ISO 26262.

Authenticity:

Degree to which the identity of a subject or resource can be proved to be the one claimed

NOTE Adapted from ISO/IEC 13335-1:2004.

The term is not used in ISO 26262, but does not mean any contradiction.

Reliability:

Degree to which a system, product or component performs specified functions under specified conditions for a specified period of time

NOTE 1: Adapted from ISO/IEC/IEEE 24765.
NOTE 2: Wear does not occur in software. Limitations in reliability are due to faults in requirements, design and implementation, or due to contextual changes.
NOTE 3: Dependability characteristics include availability and its inherent or external influencing factors, such as availability, reliability (including fault tolerance and recoverability), security (including confidentiality and integrity), maintainability, durability, and maintenance support.

The term is not used in ISO 26262, but does not mean any contradiction. But this book will address more the relation between safety and reliability.

- *Maturity: Degree to which a system, product or component meets needs for reliability under normal operation*

NOTE: The concept of maturity can also be applied to other quality characteristics to indicate the degree to which they meet required needs under normal operation.

The term is not used in ISO 26262, but does not mean any contradiction.

- *Fault tolerance: Degree to which a system, product or component operates as intended despite the presence of hardware or software faults*

NOTE Adapted from ISO/IEC/IEEE 24765.

Reliability is used in a comparable context but also for software and hardware.

The term is not used in ISO 26262, but does not mean any contradiction.

Recoverability:

Degree to which, in the event of an interruption or a failure, a product or system can recover the data directly affected and re-establish the desired state of the system.

NOTE: Following a failure, a computer system will sometimes be down for a period of time, the length of which is determined by its recoverability.

The term is not used in ISO 26262, but does not mean any contradiction.

Compliance:

Extend to which the software fulfills reliability norms and agreements

ISO 26262 compares and refers compliance especially to safety.

Usability:

Degree to which a product or system can be used by specified users to achieve specified goals with effectiveness, efficiency and satisfaction in a specified context of use

NOTE 1: Adapted from ISO 9241-210.
NOTE 2: Usability can either be specified or measured as a product quality characteristic in terms of its sub characteristics, or specified or measured directly by measures that are a subset of quality in use.

The usability of for example components describes the qualification of components in safety applications.

Efficiency:

Resources expended in relation to the accuracy and completeness with which users achieve goals [ISO 9241-11]

NOTE: Relevant resources can include time to complete the task (human resources), materials, or the financial cost of usage.

Efficiency is especially reference to the efficiency of safety mechanism in ISO 26262.

- *Time behavior: Degree to which the response and processing times and throughput rates of a product or system, when performing its functions, meet requirements*

Real-time aspects are not directly addressed, but safe tolerance time interval or any other time related requirements define constraints for safety-related functions.

- *Resource utilization: Degree to which the amounts and types of resources used by a product or system, when performing its functions meet requirements*

NOTE: Human resources are included as part of efficiency (4.1.2).

Resource usage of microcontroller is a major topic in safety engineering, but not addressed in detail in ISO 26262.

Maintainability:

Degree of effectiveness and efficiency with which a product or system can be modified by the intended maintainers

NOTE 1: Modifications can include corrections, improvements or adaptation of the software to changes in environment, and in requirements and functional specifications. Modifications include those carried out by specialized support staff, and those carried out by business or operational staff, or end users.
NOTE 2: Maintainability includes installation of updates and upgrades.
NOTE 3: Maintainability can be interpreted as either an inherent capability of the product or system to facilitate maintenance activities

ISO 262626 does not make a focus on maintainability a for example railway safety standards, but the relation between safety and maintenance is addressed.

- *Analyzability: Degree of effectiveness and efficiency with which it is possible to assess the impact on a product or system of an intended change to one or more of its parts, or to diagnose a product for deficiencies or causes of failures, or to identify parts to be modified*

NOTE: Implementation can include providing mechanisms for the product or system to analyze its own faults and provide reports prior to a failure or other event.

The term is not directly addressed, but safety analyses are key activities for element examinations of the product under development.

- *Modifiability: Degree to which a product or system can be effectively and efficiently modified without introducing defects or degrading existing product quality*

NOTE 1: Implementation includes coding, designing, documenting and verifying changes.
NOTE 2: Modularity (4.2.7.1) and analyzability (4.2.7.3) can influence modifiability.
NOTE 3: Modifiability is a combination of changeability and stability.

The term is not used in ISO 26262, but does not mean any contradiction. Especially changeability is seen more specifically in the context of a supportive process (Change management).

- *Stability: Probability of the occurrence of unexpected impacts or changes.*

The term is not used in ISO 26262, but does not mean any contradiction

- *Testability: Degree of effectiveness and efficiency with which test criteria can be established for a system, product or component and tests can be performed to determine whether those criteria have been met.*

NOTE: Adapted from ISO/IEC/IEEE 24765.
It is considered in the same context by using tests a verification measure.

- *Compliance: Extend to which the software fulfills norms and agreements in reference to changeability.*

Confirmation Measure require, questioning or examine compliance to ISO 26262.

Portability:

Degree of effectiveness and efficiency with which a system, product or component can be transferred from one hardware, software or other operational or usage environment to another

NOTE 1: Adapted from ISO/IEC/IEEE 24765.
NOTE 2: Portability can be interpreted as either an inherent capability of the product or system to facilitate porting activities, or the quality in use experienced for the goal of porting the product or system.

The term is not used in ISO 26262, but does not mean any contradiction

- *Adaptability: Degree to which a product or system can effectively and efficiently be adapted for different or evolving hardware, software or other operational or usage environments*

NOTE 1: Adaptability includes the scalability of internal capacity (e.g. screen fields, tables, transaction volumes, report formats, etc.*).*
NOTE 2: Adaptations include those carried out by specialized support staff, and those carried out by business or operational staff, or end users.
NOTE 3: If the system is to be adapted by the end user, adaptability corresponds to suitability for individualization as defined in ISO 9241-110.
The term is not used in ISO 26262, but does not mean any contradiction

- *Install ability: Degree of effectiveness and efficiency with which a product or system can be successfully installed and/or uninstalled in a specified environment*

NOTE: If the product or system is to be installed by an end user, install ability can affect the resulting functional appropriateness and operability.
The term is not used in ISO 26262, but does not mean any contradiction

- *Co-existence: Degree to which a product can perform its required functions efficiently while sharing a common environment and resources with other products, without detrimental impact on any other product*

Co-existence of function and especially software with different ASIL within common elements addresses the ability of co-existence of the different elements in common resources and its different ASIL.

- *Replace ability: degree to which a product can replace another specified software product for the same purpose in the same environment*

  NOTE 1: Replace ability of a new version of a software product is important to the user when upgrading.

NOTE 2: Replace ability can include attributes of both install ability and adaptability. The concept has been introduced as a sub characteristic of its own because of its importance.
NOTE 3: Replace ability will reduce lock-in risk: so that other software products can be used in place of the present

The term is not used in ISO 26262, but does not mean any contradiction

Those ideas and terms are illustrated in ISO 26262 in a different or similar context. For example coexistence of software of different criticality (different ASIL) doesn't see a risk if functions are similar but if these functions can influence each other negatively. Furthermore, it is important to mention that ISO 26262 uses and defines the terms validate, verify, analyze, audit, assessment and review in context of functional safety for road vehicles differently. These examples also show that requirements, terms or definitions within ISO 26262, depending from which activity or context they are used, can lead to different interpretations or meanings.

Furthermore, there are two basis process models, which need to be considered in order to observe the valid variance of processes in the development according to ISO 26262.

2.4.2 *Waterfall Model*

The waterfall model is a process model often found in the development of tools (Fig. 2.8).

This model has no specific source of origin. This is why there are so many different descriptions and interpretations as to how this model can be applied. The waterfall in general describes a higher level of abstraction than most V-models.

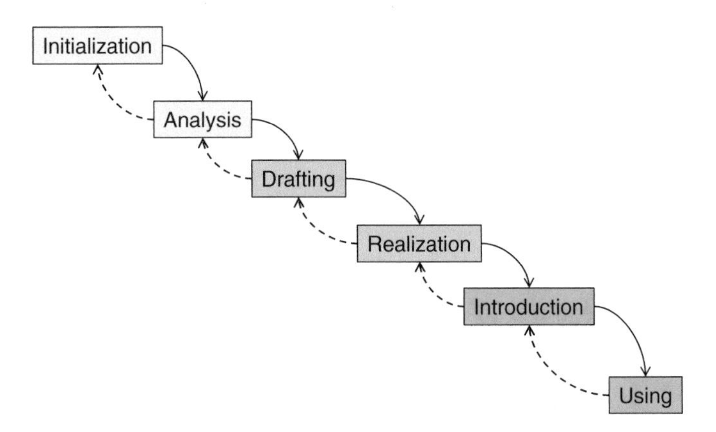

Fig. 2.8 Waterfall model [6] (*Source* Wikipedia)

Furthermore, to better picture the process one can imagine that the waterfall model transforms into a V-cycle for the design and implementation phase. Compared to waterfall models, V-based process models describe vaster parts of lifecycles. All other process models describe the initialization phase as a linear starting point that defines the interests (stakeholders, see chapter "Stakeholders of an architecture") or sources of requirements (compare to SPICE: "Requirement Elicitation") for a system.

The introduction and application right up until the product definition or the contract document and the requirement specification are often described as a linear path in process models. Iterations are not further evaluated in later phases. In addition, iterations of the planning and defining activities between the customer and the service provider are not necessarily included in the development activities.

This shows that most of the process models are derived from the IT world. A derivation of the waterfall model for the automobile industry would certainly resemble parts of the safety lifecycle of ISO 26262 or the various APQP standards.

2.4.3 *Spiral Model*

More often, the V-model is also discussed in regards to automotive industry. However, the traditional process model in this sector seems to be the spiral model.

As mentioned in the chapter 'Advanced Quality Planning', sample phases mainly determine the development activities in the automobile industry. The following figure shows the sequential process and the respective iterations in a spiral shape (Fig. 2.9):

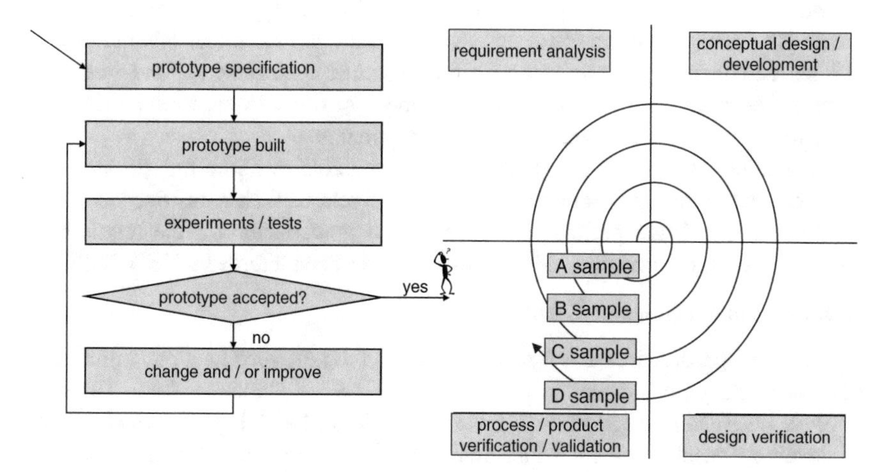

Fig. 2.9 Spiral model for a prototype- or sample-cycle approach as basis for many automotive maturity models

Today, traditional sample names such as A-, B-, C- and D-sample are only referenced in certain company standards (e.g. in Daimler's). In the APQP standards from AIAG or VDA all samples refer back to the initial sample. The sample groups for different customers are mostly aligned with the requirements of the vehicle development.

Phases of the Spiral: Prototypes

The dream of every process developer is that specification actually represents the beginning of product development. In reality, it is more an idea, which has to be built up for series production. For classic mechanics it is a trial sample that has to be complemented with essential functions or electrified for new systems. Therefore, there is often only one specification, which is defined on a higher level of abstraction in the first iterations.

Previously, in the early stage of automobile engineering, the A-sample could be made out of wood since the main focus at the beginning was the production potential of the inside of the vehicle. Today, in modern systems, the first step can already focus on the entire outer interface, so that the CAN-communication can already be adjusted to the target vehicle in the first sample delivery.

Construction of Prototypes

Since samples have to be delivered to the customers, they have to be produced in the first place. Certainly, this requires a lot of manual work in the first iterations. In the following iterations the degree of automation increases continuously. Then, the D-Sample—often comparable to the first sample/initial sample—has to be produced in the series production facility.

Experimenting—Trial/Acceptance

Moving forward, the sample has to be tested according to the given requirements. The sample will be tested in the first iterations First under laboratory conditions, and then later based on with the customer requirements, and in further stages often already in the vehicle environment in order to explore the dynamic behavior as well as the interaction of all components.

All parties involved in the process hope to be able to figure out all necessary requirements in the first shot and that the sample returns with a positive test result. In the real world, the prototype is an essential input factor for the requirement analysis. Besides simulation, this method has also been adopted in ISO 26262.

Changes, Modifications or Enhancements

Here the specifications are now changed and the new specifications introduce a new development cycle.

With DRBFM (Design Review Based on Failure Mode) Toyota was very early to develop a stable method, which introduces new iterations. Whether a specification is complete or still error-prone is difficult to examine (freely adapted from Popper: verification is positive until I can find a counterproof). Change management based of specifications can only be effective if it is known whether the specification is correct and clearly and distinctively valid for the product. Pessimists would say that this is impossible. This is why DRBFM describes a comparison based on features. In a multidisciplinary team, features are compared to functional dependencies (architecture). The positive and negative influence on a product is analyzed and assessed in a design review before proposals for modifications are accepted. This method is very useful for modern architectural developments.

The result of DRBFM is only adopted for the specification after the effects analysis and is then accepted as a modification for the product.

These aspects also influenced change management processes and requirements in ISO 26262.

2.5 Automotive and Safety Lifecycles

IEC 61508 was probably the first standard that described a safety lifecycle. The vastly simultaneously developed ISO/IEC 12207 also described a software lifecycle. It was discovered in the mid 90s that the requirements of a product could influence its design over the entire usage period. Unfortunately, it was also known that certain mistakes in all phases could lead to danger and people could get injured while dealing with certain products. ISO/IEC 12207 shows that there is a demand for the monitoring of specific error patterns of products throughout all phases of the product lifecycle. Those error patterns present further challenges for the design of a product.

The APQP standards also consider early development phases. The terms of the System-FMEA as well as later design or concept-FMEA are included in the norms. Product maintenance and the management of replacement parts have been considered by the APQP norms for quite a while. Also the idea of document archiving throughout the lifecycle has been addressed by the norms at a very early stage. The demand arose from the topic of product liability.

In IEC 61508 the lifecycle was used to define phases from the product idea up to the end of the product life, in which individual safety activities can be implemented. This lifecycle already represented the foundation to fully describe the actual requirements of a product.

Safety considerations of a product idea are already of particular importance and not only because of safety related reasons but also and most importantly economic aspects. History has shown that bad ideas sometimes can turn out to be successful. Unfortunately, bad ideas have often been pursued only because of the fear of failure and the potential hazards occurred when the possibility to prevent them no longer existed. A production stop can often cost a company more than the compensation of damages that the product might cause when used. This covers one main aspect of product liability, which has previously been addressed by the legislatures of the 19th century. For example, §823 of German civil law requests to avoid hazards of products as far as science and engineering will allow it. Also, the retailer or distributor of goods is liable for damages occurred.

Let's get back to the product and safety lifecycle. A function can cause a hazard even if it operates as intended. This is mainly referred to as safety-in-use. As mentioned in earlier chapter (safety, risk etc.) this is not addressed in ISO 26262. However, the hope is to find something throughout the course of product development that can manage the risk. Otherwise, the respective functions are limited as much as possible in order to eliminate risk.

ISO 26262 can only help to control hazards based on a malfunction of the product. Experienced engineers might be able to find safety mechanisms for dangerous functions. If those mechanisms are not found the product has no chance to establish itself on the market. It can be a real challenge to sufficiently declare such defects/faults/errors as unlikely for complex products that are produced in high quantities. Formally, the quantification of those systematic errors are not required by ISO 26262. The characteristics of such complex products, their potential errors as well as the potential variance of their usage are hard to determine. The product might still be able to enter the market but once the first hazard arises the only option is to withdraw it from sales and recall the entire vehicle. It has previously occurred that some manufacturers have had to buy back vehicles. This is why one of the first steps in order to get to the field of application of ISO 2626 is to prove the safety of use/usage safety of the product. In order to avoid potential liability issues, it is useful to clearly document all safety issues to prevent safety-in-use being questioned after certain changes are made during the following development. Generally, the nature of an engineer is not to scrap an idea after the first failure but to adapt and modify it accordingly.

In order to take a brief look at the end of the product lifecycle, let's discuss certain aspects of the product lifecycle itself. In regard to hazards, the public discussion on mobile phones would be a good example.

Of course, a lot of qualitative electrical waste is produced due to the fast and short lifecycles of electronic products. This wouldn't be questionable from a safety perspective if the components themselves were not so expensive and were not environmentally damaging materials, such as lead. This is why the government implemented clear procedure rules. Now it may be far-fetched to say that the toxic electrolytes in capacitors may also eventually cause environmental damage or that the burst of an electrolyte capacitor is a malfunction of an E/E element.

But the question here is whether or not ISO 26262 is helpful in this regard. In fact, there is a potential for hazards that have to be considered in the production and development of products, to prevent issues with product liability. In cases such as these, it is important to consider the possible end of a sub product. In general, cars are used beyond their actual warranty. Cars that are over 25 years old can possibly become classic/vintage cars and more popular than a car with the latest engineering technology. Luckily, cars made 25 years ago used far fewer electronics. However, this will now change from year to year. It is particularly important to consider the maintenance of the car, particularly those components and systems that are subject to wear and tear.

Opel once advertised with a lifelong warranty, meaning 15 years and 160,000 km (99,419.3908 miles)—a campaign that was quickly abolished. We also learned that NASA bought Intel's 8086 microcontrollers via eBay in order to be able to maintain old systems. It is increasingly difficult nowadays to maintain program parts written in FORTRAN. Such dated systems are practically impossible to re-lay with systems such as WINDOWS and other advancing computer systems. Going forward, we will see that a prognosis for an electrical component beyond the failure mode of more than 10 years is extremely difficult. Nowadays, in the field of utility, vehicles lifespans of over 20 years are projected. Intermittent errors have already been detected in an 8086 but it is highly questionable whether actual measures in the integration have been undertaken.

To assure maintenance according to safety aspects will become a real challenge for the automobile industry.

2.5.1 Safety Lifecycles for the Development of Automotive Products

ISO 26262 describes safety lifecycles in part 2, Chap. 5 "Overall Safety Management". Here, the idea is to inter-relate safety lifecycles, product lifecycles and the "Management of Functional Safety". The aim of the management of functional safety according to ISO 26262 is to define the responsibility of acting individuals, departments and organizations that are responsible for each individual phase of the safety lifecycle. This applies to necessary activities, functional safety for products, and the vehicle system—or as referred to in the norm—the ITEM, as well as measures must be taken in order to confirm that the products are developed according to ISO 26262 guidelines.

Moreover, other activities have to be described that are necessary and important beyond the safety lifecycle in order to show a respective and appropriate infrastructure in order to apply the product lifecycle. Very important here is an applied and utilized quality management system and safety culture to ensure that each individual employee, right up to the top management regard safety with the required diligence and respect in order to implement and apply the necessary measures appropriately. Further crucial premises are the systematic learning process from previous mistakes, competence management and continuous improvement such as qualification and training programs in order to apply a safety lifecycle.

ISO 26262 generally assumes that products are developed within a project structure. Here there is a chance that divisions or organizations develop products according to a general interpretation or implementation of a product lifecycle ("Project independent tailoring of the safety-lifecycle"). This means a process scope is developed, that represents a valid derivation of ISO 26262 but can also be optimized in regards to infrastructure and product aspects.

Alternatively, each product development can be directly derived from the scope of ISO 26262 as for example as project safety plans. Especially in product development and production it can be favorable to define many activities—customer and/or product alike. This can be of advantage in machinery utilization or in the scope of product development. Internal processes can be coordinated and aligned appropriately to qualified development tools and different versions for various customers can be offered with little effort. Also, the reuse of established processes, safety components or products can have a positive effect on the safety of all products.

2.5.2 Safety-Lifecycles According to ISO 26262

The safety-lifecycle of ISO 26262 summarizes the most important safety activities in the conceptual phase, the series production and the series production release. A central management task is the planning, coordination and proof of these activities throughout all phases of the lifecycle. Volumes 3, 4 and 7 describe the activities of the conceptual phase, the series production and those according to SOP thoroughly (Fig. 2.10).

This safety-lifecycle directly refers to the respective chapter in ISO 26262. The management of functional safety according to part 2 of the norm includes all further activities from part 3 (Concept Phase) to part 7, Chap. 6 (Operation, service (maintenance and repair), and decommissioning)

The safety-lifecycle is divided into 3 phases:

- Concept
- Product development
- After production release/approval

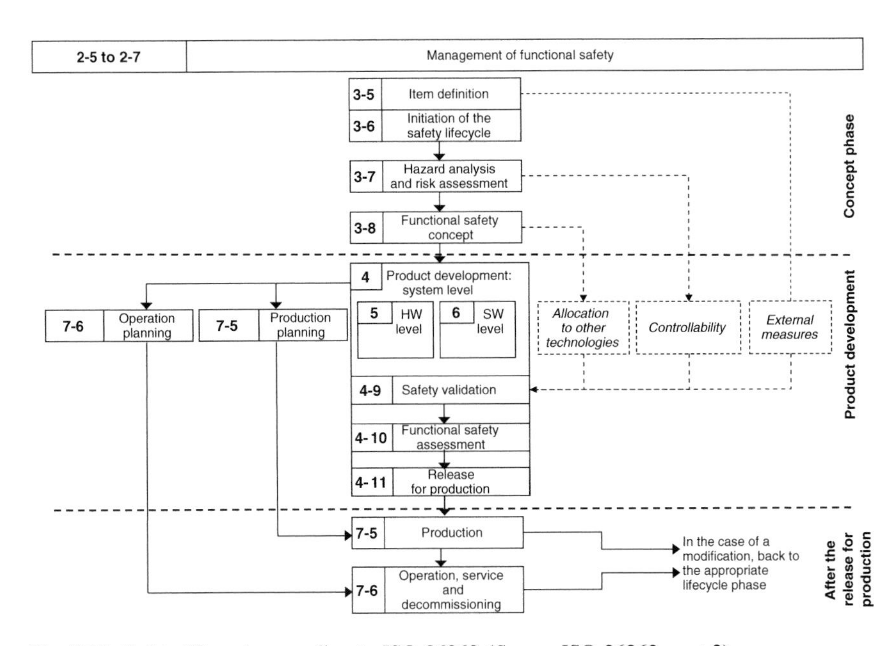

Fig. 2.10 Safety-lifecycle according to ISO 26262 (*Source* ISO 26262, part 2)

Please note that the technical safety concept is associated with the product development. Next to the 3 parts of product development of systems, EE-hardware and software, and the chapters about production development and plant engineering (part 7) are described. Those are activities that are considered besides the development V-cycles. Furthermore, some activities are mentioned that are not directly addressed by the norm but often necessary for the product development.

External Measures
These are measures that are not influenced by the observation unit, which are described in the system definitions. External risk reduction includes for example the behavior of road users or characteristics of the road itself. This is also described in the system definitions. External risk reduction is seen as profitable within the scope of the hazard and risk analysis. The proof of efficiency of external risk reduction is not included in this norm.

Controllability
Controllability, the underlying concept of the hazard and risk analysis, should be proven within the phase of product development. If it does not relate to the distinct controllability of individuals exposed to hazard, then it is covered in part 3 of ISO 26262. This part overlaps with the content of safety-in-use, since the question whether functions are defined in a way that they are not dangerous when functioning properly is also relevant.

Association to measures of other technologies
These are technologies that are not covered in the scope of this norm, for example, mechanics and hydraulics. They are addressed when associated to safety functions. Also, the proof of efficiency or effectiveness and even the application of these measures are not part of this norm.

In the scope of the functional safety management the norm requires certain activities for the safety-lifecycle:

- Sufficient information has to be documented to the E/E-system for each phase of the safety-lifecycle, this is necessary for the effective fulfillment of the following phases and verification activities.
- Management of functional safety has a duty to ensure the execution and documentation of phases and activities of the entire lifecycle and to provide a corporate culture that promotes functional safety.

From the point of view of functional safety it is not about the fulfillment of the requirements that are derived from any process models. The safety-lifecycle has to be derived correctly and sufficiently. It is important for project planning and the planning of safety activities that the safety concepts are implemented in a way that sufficiently ensures safety goals.

2.5.3 *Security-Versus Safety Lifecycles*

For meaningful safety-related product development not any quality characteristics could apply their own process. Therefor also even if there are other means of analysis or methods for verification or validation necessary, it is a matter of tailoring of the product lie-cycle to apply activities to as necessary for all non-functional requirements also such as security. Similar to challenges with the safety lifecycle for safety-related active safety functions and other passive safety functions the tailoring and even the entry into the safety lifecycle is different. The intended safety function for an active safety function should be made safe by adequate measures during the Item Definition, and for typical passive safety functions it should be done during entire safety lifecycle.

Mayor security threads are categorized as follow:

- Availability
  Assures access to data and infrastructure
- Integrity
  Identification of manipulation of data on controller or communications
- Confidentiality
  No unauthorized information access

A particular security topic is theft-protection, since this provides many dependent functions to Functional Safety.

Furthermore, all “Integrity” related issues are very often also causes for “Functional Safety” impacts.

References

1. [ISO 26262]. ISO 26262 (2011): Road vehicles – Functional safety. International Organization for Standardization, Geneva, Switzerland.

   ISO 26262, part 3, appendix B1: 11

   ISO 26262 Part 2, Clause 5.3.2: 17

   ISO 26262 Part 2, Clause 5.4.4: 18

2. [IEC 61508]. IEC 61508 (2010): Functional safety of electrical/electronic/programmable electronic safety-related systems. International Electrotechnical Commission, Geneva, Switzerland.

   IEC 61508, Part 1, Part 5, A5: 9

3. [ISO TS 16949]. ISO/TS 16949 (2009): Systems. Particular Requirements for Application of ISO 9001:2008 for Series- or Spare parts Production in Automobile industry; VDA, 3rd English edition 2009.

ISO TS 16949, 4.2.3.1:	13
ISO TS 16949, 5.6.1.1	13
ISO TS 16949, 5.6.2	14
ISO TS 16949, 5.6.2.1:	14
ISO TS 16949, 5.6.3:	14
ISO TS 16949, 6	15
ISO TS 16949, 7.3.1.1:	15
ISO TS 16949, 7.3.2.3:	15
ISO TS 16949, 7.3.3.1:	16
ISO TS 16949, 7.3.3.2:	16
ISO TS 16949, 7.5.1.1:	16

4. [VDA FMEA] VDA (2008), Volume 4 Chapter, Product and Process FMEA, QMC, Berlin 20
5. [ISO/IEC 25000]: ISO/IEC 25001:2007, Software engineering—Software product Quality Requirements and Evaluation (SQuaRE) — Planning and management 25
6. [waterfall model]: Figure 2.8: Waterfall Model (Source: Wikipedia) 30
7. [unpublished research project], for further information available
8. [V-model XT 1.2], V-Modell® XT, Version 1.2.1.1, IABG, 2008
9. [APQP AIAG], APQP AIAG 4th Edition, Automotive Industry Action Group, APQP, 2006

Chapter 3
System Engineering

The chapter tries to give answers on:

- What is system engineering?
- What is the relation to system safety engineering?
- How does it differ from the requirements of the automotive industry in other domains?
- What does it have to do with Functional Safety?
- What is the necessary impact on organizations?

General process models often fail to answer the questions, how to enter the V-, the spiral- or the waterfall-process model. Which aspects need to be considered so that activities within a V-cycle can be planned and intermediate targets defined or maturity of quality factors could be considered? The aim is to show a general approach or procedures that can be considered in each phase of the product development or represent the foundation for development activities in general.

3.1 Historic and Philosophic Background

A skeptical mind state is nothing new to human nature. Dating as far back as Socrates people were often cynical about things they saw and heard, leading to some scholars conclusions at times being questioned.

When did questioning technical coherences start?—In Greece 600 before Christ? Did it start in Egypt and the inhabitants of Mesopotamia just didn't document it?—It remains undecided. However, since written documents have existed, people have tried to describe certain phenomena and draw their conclusions from it. Ionic philosophers, as for example Pythagoras, did this in a very mathematical way. He certainly did not imagine that his formula would one day be used to energize an engine in order to bring blind power and active power in relationship.

H.-L. Ross, *Functional Safety for Road Vehicles*,
DOI 10.1007/978-3-319-33361-8_3

It is said that in the school of Elea, Parmenides taught that you can observe things but you cannot draw arbitrary conclusions from it. He questioned whether the observed allows for conclusions to be drawn even before Socrates. Up until today, 2600 years later, we struggle with the question whether the reason for a negative test result is a wrong test hypothesis or the test wasn't appropriate in order to make a qualified statement if requirements were implemented correctly and sufficiently. Democritus tried to define the term 'atom' as the smallest element or building block of everything that exists but Nils Bohr discovered that there were even smaller elements than atoms.

Albert Einstein was not the only one who showed that there are various forms of interactions and that we need different models in order to describe the observed. Aristotle knew: "The whole is more than the sum of its parts." Not only the elements and their characteristics determine how the elements react to each other, it is also important what environment the elements interact. Today we know that the stability of a screw and a dowel is different in a plasterboard wall than a brick wall. The stability also depends on the design of the walls. Despite observing and drawing conclusions, Aristotle also raised the topic of induction. The fact that the entire mathematical induction is now described as a deductive method shows that words are subject to change throughout the years. This isn't the only example where mankind had to re-learn something from scratch that had already been discovered thousands of years ago. In the beginning of the 13th century, Roger Bacon described methods for an electric engine while studying magnetism. The idea of a "continuously moving wheel" leads to the realization that a "perpetuum mobile" does not exist. Werner von Siemens apparently did not know the work of Roger Bacon.

In more recent times, Karl Raimund Popper described our dilemma nowadays saying that it is not possible to verify something—only to falsify specific characteristics. Here the example of the swan is often mentioned: Mankind always believed that only white swans exist until the discovery of black swans in Australia. How do we deal with that? The word 'verifying' comes from the two Latin words "veritas" (truth) and "facere" (to do). They say "In vino veritas"—"In wine (there) is truth", but what truth, one can only try to guess the next day after consuming too much of it. Apparently, different people use the word verifying in different ways. Popper left numerous clues for falsification. According to him, if a result is negative, we are unable to question the entire statement or hypothesis. However, he encourages us to use this test result to derive new insights, which give clues as to what needs to be changed in order to make the test result positive. Even if all test results are positive, we still haven't fulfilled all requirements yet. The lesson we learn here is that we should analyze negative test results in order to find ways to improve the product. This leads us to the conception that the basic principle of proof of safety in today's safety standards can be seen as follows:

> If all devisable mistakes/errors/faults of a system are brought under control, the system is regarded to be safe.

This perception can be problematic for new developments or if new technologies replace traditional ones for established and proven vehicle systems. This pertains to the entire "By wire" systems but particularly for remotely controlled systems, which so far have been operated solely by the driver. The guideline here is: "Equivalent Level of Safety", which means that for example new electronic systems need to be as safe as the conventional hydraulic system.

If an established and proven system is implemented according to comparable principles, it is sufficient to show the compliance of safety principles. New products in a new technology require a systematic proof of safety. These new principles and guidelines are, besides the norms, mainly included in all international, worldwide and industry wide standards.

This was a small digression into philosophy but it is also important to refer to certain engineers, physicists and mathematicians in this chapter.

George Boole (1815–1864) is considered to be the inventor of Boolean algebra. The rules were generally known before his time but he was the one to formulate them in his book "An investigation of the law of thoughts" as "logical algebra". Augustus DeMorgan formulated the DeMorgan's law, which influences the deductive analysis. ISO 26262 [6] also includes the quantitative safety analysis besides the qualitative inductive and deductive analysis. Here it is necessary to also mention the names of people who developed the essential basics of safety engineering.

Before (or during) Second World War, Robert Lusser formulated his "Conformity of reliability chains" and Erich Pieruschka amended the quantification. Those two gentlemen probably knew Russian Kolmogorov or at least the German version of his book in 1933 "Foundations of the Theory of Probability". The Axiom of Kolmogorov says: "The probability of a combination of a countable sequence of disjoint events equals the sum of the probabilities of each individual event" in a slightly shortened version. The beta error derives from the Kolmogorow-Smirnow test. Beta errors or beta factors are used in safety engineering to describe dependencies.

Furthermore, it is important to mention Andrei Andrejewitsch Markow, whose models where not only essential for the speech recognition but who also taught us how to quantify transitions of different conditions.

This historical digression should show that we are not reinventing the wheel with functional safety but that we aim to describe and analyze technical systems. For this purpose we use historically proven and established methods of safety engineering and refine and enhance them.

3.2 Reliability Engineering

The first researches in reliability engineering in connection with today's mathematical term started at the beginning of the industrial age. A complete study of the lifespan of a roller bearing was documented as part of a technical railway development. The law of Robert Lusser describes a chain of elements, where the total reliability occurs from the product of the individual reliabilities. This describes the foundation for reliability of all technical systems. Basically, this law says that: "The chain is as strong as its weakest link". Also safety related functions or safety mechanisms can only work as well as the individual parts of which they consist of. This is why the partition and structuration of mechanisms of action is the essential task of a failure or safety analysis. It is important to analyze the identification of the demand of additional mechanisms and the intensity with which they influence the system. This analysis shows the appropriate measures to make the system more reliable, less susceptible to maintenance and reach a higher level of safety and availability in order to strengthen the chain in its weakest parts. Destructive tests, in which a stimulus is implemented or injected in the product, exist and they change the analysis of the product as well as the purpose of the analysis. However, the analysis itself does not change the product, only the measures, which are added or develop a new behavior or changes in characteristics through a modification (for example in the context of a process of change) in the product, are the aim of an analysis.

Until about 1930 the activities in the field of reliability were mainly limited to mechanical systems. The focus of the efforts for electrical systems was to ensure electric energy sources, which means to raise the availability of such. Parallel electrical gearshifts from transformers and transfer units, meaning the insertion of redundancies, were a substantial progress in the electrical reliability.

New concepts also developed in the field of aeronautics/aviation that considered reliability engineering. For example, observing the failure mode of various aero structures through the determination and evaluation of statistical data. Especially the insertion of redundancies helped to maintain functionality. Also, technical availability and the possible measures to increase it were developed systematically. The largely qualitative signal chain analysis became then also quantifiable through statistic considerations developed by the team of mathematician Erich Peruschka. He defined the following principles:

R1, R2, … Rn are survival probabilities of the individual chain links.

Since all links of are required for a chain to function and the survival probability of each individual link is independent of each other, the survival probability of the entire chain as the product of its individual probabilities is calculated according to the rules of the probability theory as follows:

Total survival probability of the chain: **Rg = R1 · R2 · … · Rn**

Consequently, the reliability of individual components of a system exceeds considerably the reliability of the total system. A new and predominantly technically oriented discipline: The reliability engineering. This discipline addresses the measurement, prediction, maintenance and optimization of the reliability of technical systems. Reliability engineering boomed in the '50s in the United States through the growing complexity of electronic systems, particularly in the military. The analysis of failures and their cause as well as the reparation of defect components became increasingly time, money and resource consuming. Hence why the US Defense Department founded the *Advisory Group on Reliability of Electronic Equipment (AGREE) in 1952*.

Researches showed that the maintenance costs for electric systems were twice as high as the procurement costs. This led to the insight, which reliability engineering has to be an integral part of development and construction. AGREE insisted that new systems and components have to be extensively tested under extremely harsh conditions (temperature, voltages, vibrations etc.) in order to discover and correct blind spots (weaknesses) in the construction. Furthermore, it was recommended to calculate the mean time between failures (MTBF) and its confidence interval. Also, it must be proven that the mean time between failures is above the required value. Reliability engineering was also adopted for electrical components and from thereon within the next 20 years practically for all other technical domains.

The necessity to control reliability of technical products is nowadays extremely important because of the increasingly competitive environment and the consequent market pressure, which is no longer only controllable through price. The rapidly growing technological progress leads to shorter product cycles. The time pressure does not allow for extensive practical testing before the product launch. The rising cost pressure requires inexpensive development and manufacturing techniques, which must not influence the quality of reliability of the product. All these factors increase the risks of product development. Reliability engineering provides methods for risk limitation in conception, development and production of technical products.

Reliability is one aspect of technical uncertainty. The reliability method addresses the prediction, measurement, optimization and maintenance of the reliability of technical systems. This requires the application of statistical and probability theoretical methods. Probability is the only to predict whether or not a product can continue to function for a specific period of time.

Reliability is technically seen a quality factor. Very important in this context is the necessity to explicitly define the desired function of a product and thus its functionality. Reliability on the other hand, is only quantitatively describable if connected and referred to, through time. Reliability engineering helps to make observations and statements of the behavior of a system during its utilization period. However, the reliability of a technical product does not solely rely on the utilization period. The usage frequency and intensity as well as the environment in which the system is used also have an important influence on reliability. Consequently the environment and the usage profile also play a major role in reliability.

3.2.1 Foundation/Basis of Reliability

Reliability is generally described as the expected function fulfillment throughout a defined time period. The mean time between failures (MTBF) is the classical measure for the reliability of components (for example devices, assemblies, equipment, facilities). Here we differentiate between reparable and non-reparable components. The mean time to failure (MTTF) is defined for non-reparable components; if needs be, the mean time between failure (MTBF) can be referred back to through various maintenance models. The mean time to failure, MTTF, determines the statistical expectation period of a failure. According to IEC 60050 MTTF is defined as follows: “The expectation of the time to failure”. For a lifespan allocation with a constant failure rate (mostly similar to an exponential distribution) the reciprocal of MTTF is the failure rate (R). In FIT (Failure in Time) the failure rate is stated as the entity “failure per 10E-9 h” and it is also a measure for the failure probability of electrical components in safety engineering.

$$\mathrm{MTTF} = 1/\mathrm{R}$$

Typically for the automobile industry is an average statistical life expectation of 15 years with an annually driving performance of 300 h. If one failure in the life span of a vehicle occurred out of one million components, the failure rate would be:

MTTF at continuous demand of an action throughout a lifespan: MTTF = 15 years × 1 Million components/1 failure = 15,000,000 years MTTF at an effect of an action solely during travel time of a lifespan:

$$\mathrm{MTTF} = 15 \times 300\,\mathrm{h} \times 1\,\mathrm{million\ components}/1\,\mathrm{failure} = 1,500,000,000\,\mathrm{h}$$

Failure rate as reciprocal of MTTF in FIT:

$$\mathrm{R} = 1/1.5\mathrm{h} \cdot 10\mathrm{E} - 9 = 0.67\,\mathrm{FIT}.$$

A continuous demand in action would lead to a different value for the failure rate because the time span for the effect in action is considerably bigger. The lesson here is that not even the failure rate represents a distinctive and definite value for a component.

The type of operation and the operational environment have to be taken into account for the determination of values. Whether the repair time is added into the models depends heavily on the operating conditions. The automobile industry usually uses the expected lifespan, which can sometimes be extremely challenging. A cable harness in a car is usually listed as 6000 Fit, which means it is always the weakest link of the chain. We often tend to search for the potential for optimization of a chain at this place. In a quantitative observation of functions, this high error rate for the weak link would dominate all other errors in a signal chain so that the quantification would not show the desired effects at this point. A lot of effort is put into the improvement of cable routing and connections. Very frequently errors arise in the cable harness due to heterogeneous usage, the fitting style in vehicles or improper maintenance. This is why such connections are often only shown as mere formal with a FIT in the observation of safety applications. This is frequently recommended by most of the manuals for reliability. In reality there is no constant error rate throughout the entire lifespan because a continuous and steady effect of an action hardly exists. A conservative design can often prevent components being used beyond their elastic limits. A statistically spread aging behavior is no longer reasonable if these limits are often exceeded. No material has a constant aging curve and also a material diversification, depending on the effect of actions this leads to differences in the aging behavior.

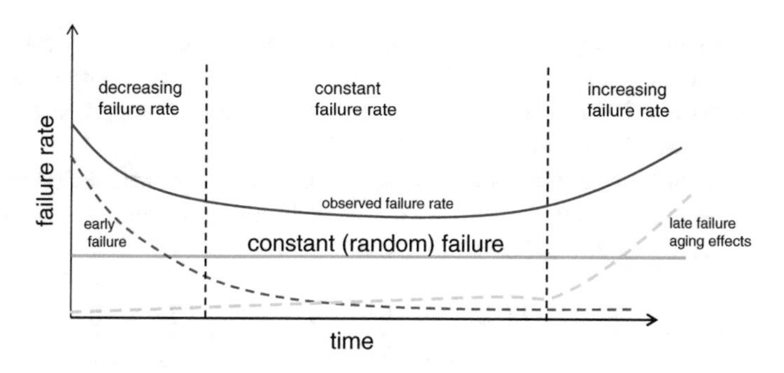

Fig. 3.1 Bathtub curve (demography example)

This fact led to the definition of the bathtub curve, which is often used as a reference model in statistics. It facilitates the observation and its extent and sufficiently offsets variances especially for electric components (Fig. 3.1).

The bathtub curve shows three areas over time. The early failure phase describes the time frame in which the failure behavior is not sufficiently developed through unknown influences, environment parameters, correct materials and bias points. This should be investigated for the development of components within the context of the design verification so that phase 2, the usage phase, can be entered at the beginning of series production. The usage phase should be designed in a way that the failure rate only starts after the expiration of the statistical life expectation of the components. In reality the failure rate is placed below the bathtub curve, as far as necessary so that an age induced increase can be seen and a sufficient robustness level ensures that the statistical life expectation is achieved. ISO 26262 does not mention any requirements, for example in order to prevent early failure behavior.

So called 'Pi-factors' are used in order to try to standardize and adjust or correct environmental conditions. Typically, Pi-factors orientate themselves at the Arrhenius equation.

$$k = A \cdot e^{\frac{-E_A}{R \cdot T}}$$

A pre exponential factor or frequency factor
E_A activation energy (entity: J mol^{-1})
R = 8314 J K^{-1} mol^{-1} universal gas constant
T absolute (thermo dynamic) temperature (entity: K)
k reaction speed constant

The following formula is used if a temperature dependency for A exists:

$$k = B \cdot T^n \cdot e^{-\frac{E_A}{R \cdot T}}$$

However, particularly for electric components proven handbook data are used as reference, since the bathtub curve and the material dependency in such formulas already represent a strong abstraction of the measurable result of technical systems. One of the most common handbooks for reliability of electric components is the Siemens norm *SN 29500* [2]. This norm describes a simple approach to manage correction and was later added to *DIN EN 61709* [1].

The temperature dependent acceleration factor π_T for 2 failure mechanisms (e.g. for discrete semiconductor devices, IC's, optoelectronic components...) are stated in DIN EN 61709 or SN 29500 as follows:

$$\pi_T = \frac{A \times EXP(E_{a1} \times Z) + (1 - A) \times EXP(E_{a2} \times Z)}{A \times EXP(E_{a1} \times Z_{ref}) + (1 - A) \times EXP(E_{a2} \times Z_{ref})}$$

If A = 1 and $E_{a2} = 0$ the above mentioned relation can be referred back to the basic model for failure mechanisms (e.g. for resistors, capacitors, inductance) described in Sect. 3.3.

$$\pi_U = EXP\{C_1 \times (U^{C2} - U_{ref}^{C2})\}$$
$$\text{oder}$$
$$\pi_U = EXP\left\{C_3 \times \left[(U/U_{rat})^{C2} - (U_{ref}/U_{rat})^{C2}\right]\right\}$$

Stress factors or voltage/tension dependencies π_U according to DIN EN 61709/SN 29500,

or

$$\pi_I = EXP\left\{C_4 \times \left[(I/I_{rat})^{C5} - (I_{ref}/I_{rat})^{C5}\right]\right\}$$

Stress factors for electricity dependencies π_I according to DIN EN 61709/SN 29500.

Besides the environmental factors, the kind of failure distribution and how it can be statistically described for the various different technical elements, also play an important role.

The most popular distribution is the normal distribution or Gaussian distribution. The Gaussian bell curve used to be printed on 10 German Mark bills (Fig. 3.2).

Fig. 3.2 Normal distribution or Gaussian bell curve

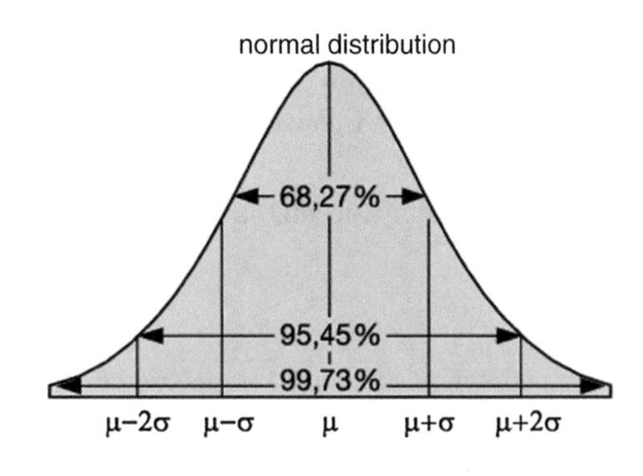

Production engineering often considers the value of 6 Sigma (six sigma). 6 Sigma considers 3.4 defects per one million failure possibilities, a failure probability of 0.00034 %, an absence of failure of 99.99966 % within a reference period or also a short-term process capability of Cpk = 2 or long-term of Cpk = 1.5. Countable values, based on natural numbers of electric components, are often called Chi-Square distributions and binominal distributions, logarithmic or Weibull distributions are also often considered for failure probability.

In the automobile industry, AEC (Q) 100 is used for complex components. It is a standard for the qualification of electric components. Simple components as resistors or capacitors are not covered in this standard. Since these simple components would often push all statistic boundaries through their variety of elements, such statistic observations are often insufficient for safety engineering. The risk for such simple components is that harmful components can be delivered to the production undetected. This is why the eligibility and whether the components are actually sufficiently dimensioned for their case of application are tested in the context of the qualification of the entire electric assembly group. The value for failure rates is taken from the reliability handbooks. However, for the correct qualification including the proof of lifetime efficiency of the entire electronic assembly group it is assumed that the simple components is within the constant phase of failure rates of the bathtub curve.

3.2.2 Reliability and Safety

Reliability is generally described as a component characteristic as opposed to safety, which is seen as a system characteristic. Principally the rule is that reliability is only considered as a component characteristic if the environmental conditions are clearly defined. The question is whether the same challenges regarding reliability for components in complex dynamic systems apply to safety.

First of all the question arises if a component environment is entirely specifiable. For many, especially solely mechanical components, we can assume normative and always consistent environmental conditions. But if we observe reliability over time, we find influential factors that are hard to specify or often only known as a result of negative experiences. This applies for material compatibility for the materials cupper or zinc or stainless steel and salt environment, which can turn into a galvanic element at a certain concentration. As a consequence it could lead to corrosion and other chemical reactions.

Furthermore we know that shock, strokes or friction, at varying strengths, surface character and material combinations can lead to more or less material deterioration and even cracks in the material. Also the intensity or impulse with which components interact plays a big role in the lifetime reliability. Some materials consider a blow or stroke of certain strength (and also a specific amount per time unit) as an elastic blow so that it does not cause significant aging effects (meaning the material or its structure remains unchanged). There are also certain changes in the interaction of materials involved. This can depend on dirt, humidity or other chemical substances. A significant example is that the comparison of a strength that hydraulically influences a component is often considered to be a soft impulse because the hydraulic liquid itself absorbs shock and the build-up of pressure in hydraulic is often softer. If the strength is of mere mechanic nature, maybe even based on an electromotive strength, the impulse for the components can be considerably harder. This can have a crucial influence on the firmness requirements up to the lifetime reliability. There are different connecting starting points for functional safety, where these two topics overlap, for example all external or outside intersections and the component intersections.

The definition of a vehicle system (ITEM definition, ISO 26262, part 3, chapter 5) already specifies external measures, environmental conditions, behavior with external vehicle systems, operating conditions, dynamic behavior etc. This means that essential influential factors of reliability and safety have to already be considered for the goals for functionality and the technical intersections of the vehicle. Such parameters should be inputs for the hazard and risk analysis. In this regard, we will find different results in the observation of potential malfunctions, which can lead to danger, especially for the parameters S (degree of severity) and C (Controllability through the driver (or other people involved)). An example would be the predetermined breaking point of the transmission, which should prevent a blockade of wheels. Blocking transmission could lead to a blockage of the entire powertrain. Of course, such a break-off must not happen for an appropriate load and has to be guaranteed throughout the entire usage period and lifespan. The switching time of modern transmissions become shorter and shorter in order to minimize energy losses and to achieve better speedup results. Because of that gears are shifted harder, which means that the impulse and the energy used is much higher. As a result, the predetermined breaking point of the transmission can no longer be interpreted over the lifespan and we are forced to introduce E/E-measures against the blockade of the power train.

This measure can quickly lead to a high ASIL because rear axles are responsible for stabilized driving of the vehicle.

The aim of system development is mainly to describe the design in a neutral way, so that the reliability first and foremost becomes relevant in the design of the components. Software design often discusses reliability but ISO 26262 does not formulate concrete requirements for systematic methods in order to determine the reliability of software. However, for mechanic components the same statements apply as those formulated for the vehicle system and its integration into the vehicle.

For the electronic we often find very tight intersections, particularly because the evaluation of the hardware architectural metrics (ISO 26262, part 5, chapter 8) as well as the evaluation of safety goal violations due to random hardware failures (ISO 26262, part 5, chapter 9) are based on the failure probability of electric components or occurrence probability of random hardware errors. Section 4.4.2.5 covers such quantitative safety analyses in detail. Often, chapter 7 part 5 of ISO 26262 is overlooked, which covers the correct dimensioning and verification of

EE-hardware design according to safety requirements. It also derives the respective electronic design and safety requirements from the safety and system design.

Of course, a resistor has certain reliability in a normative environment—but does the basis of reliability handbook data really reflect the actual environment of components? If implemented safety mechanisms are not considered, mere reliability prognoses can be made for the individual components in the context of the design verification. These values also represent basis of the respective quantitative metrics of functional safety. Another aspect is most likely the probability with which a specific error repeats itself in the design. For a mere functional observation there are often no clues for such dependencies. However, if we consider the realization, the size, the electric resistor, distances on the printed circuit board, diameter of the conductor line or connector pins, materials and material compatibility, thermos conductivity etc. can have an essential influence on the reliability and eventually also on the safety of products. The dimensioning of electronics becomes an important safety measure if we look at dangerous heat developments up until fire as a potential error function or the electronic that needs to be realized. For the most part this leads to design boundaries for the total product. This means that the safety margin will have a significant influence on the possible performance of the final product.

3.3 Architecture Development

Architecture is often seen as the spine of each product. ISO 26262, part 1, chapter 1.3 describes architecture as the representation of a vehicle system, of functions, systems or elements, which are identifiable through components, their distinctions, intersections and allocation to electric hardware and software. The functional concept (ISO 26262, part 1, chapter 1.50) is mentioned as the basis for the definition of vehicle systems. According to the glossary the functional concept is compiled from specifications of intended functions and their interactions in order to achieve the desired behavior. Therefore, it is evident that architecture needs to fulfill two requirements. It provides the product structure and its intersections as well as the foundation for the description of the technical behavior. Each component or element and their intersections ask for certain requirements. The intended behavior as well as the behavior in case of a failure has to be specified. This forces us to plan and define all levels of abstraction, perspectives, intersections as well as their desired technical behavior in advance. Originally, the term safety architecture is defined as a further term of architecture in ISO 26262. However, no clear distinction from product architecture could be agreed upon. Particularly the intersections and interfaces of the product must be defined consistently for the safety relevant parts as well as all other parts of the product. Furthermore, some argue that in this matter architecture is actually referred to as safety architecture. However, this term would not comply with the functional concept idea since it would give the impression that all parts and characteristics that are important for the realization of a safety related product are automatically safety relevant themselves. Generally this needs to be avoided. Safety

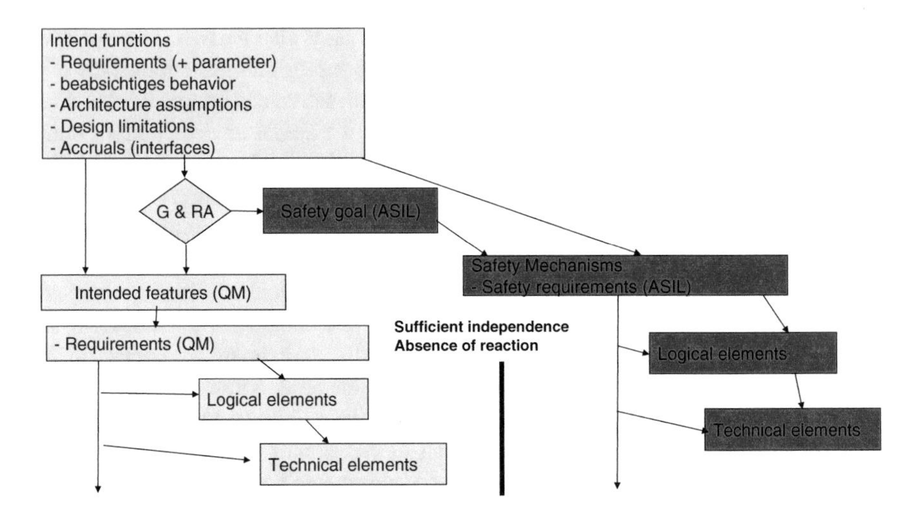

Fig. 3.3 Example for requirement management; Separation of intended functions (QM) and safety mechanisms (ASIL)

mechanisms and safety relevant functions should be easy and clearly defined even if the functions and their interactions become extremely complex.

Figure 3.3 shows how such a structure can be planned but also how requirements from the definition of the vehicle system strongly influence all elements of the architecture. If for example the intended function is a non-safety relevant function (QM) and implemented in the same technical element (e.g. microcontroller) each characteristic of the microcontroller can also influence safety relevant functions. This is why older safety norms mention that all functions in a microcontroller need to be implemented in accordance to the highest safety integrity level (in this case ASIL). This was possible to be achieved for systems with only one safety goal through separate microcontrollers. However, it was extremely difficult for systems with multiple safety goals, various levels of safety integrity or ASILs and different safety conditions. This is why it is important to analyze the entire product architecture and the respective integration environment as a whole. ISO 26262 also includes and allows the possibility that one vehicle system can consist out of several systems. If in this case the intersections of the individual systems are not well matched and aligned they will have to be adjusted in the integration process. Otherwise, there will be no systematic alignment of intersections. Therefore, if they are not planned beforehand, architecture and the various systems will define their own intersections. It would be sheer coincidence if they would match the respective systems or the intersections of the vehicle, into which each system has to be integrated.

In regards to the vehicle, an electric system includes the physical identification of sensors and the actuator providing the vehicle reaction. ISO 26262 considers this as a function or functionality of an electrical system. The same applies for the software—it can be described as a component of a microcontroller or alternatively, the functional behavior including the microcontroller can be described as a system function and the software or multiple software components and the microcontroller as two or more components of which the system is composed from. The commitment to one definition implies that by definition excluded solutions, if mistakenly implemented, can cause new technical risks. If no interrupts are used in a microcontroller there are no risks that they can cause an error. An old (VW) Beetle had no electric apart from the transistor radio and therefore no safety related risks based on failures of electric components.

3.3.1 Stakeholder of Architectures

What is the purpose and aim of architecture and what people, groups or (in order to stick to process descriptions) what role does architecture need to have? What is the difference between a stakeholder of a system or a product? Generally speaking we should say 'stakeholder of the product' but in this case we limit it to architecture.

Figure 3.4 shows the driving forces of architecture, meaning that all these aspects can influence the requirements for a product. Here it is important to identify the driving forces of a product development and in a negative sense, the risks for a product, at a very early stage—especially when considering the cost factor. However, almost all driving forces can have analogical influences and risks for various developments.

Development requires money for the development resources, tools, laboratory equipment, production goods etc. Even big and popular inventors in the automobile industry such as Benz, Diesel, Otto etc. were familiar with this issue. Money is the reason why we define certain characteristics very strictly and search for cheap and

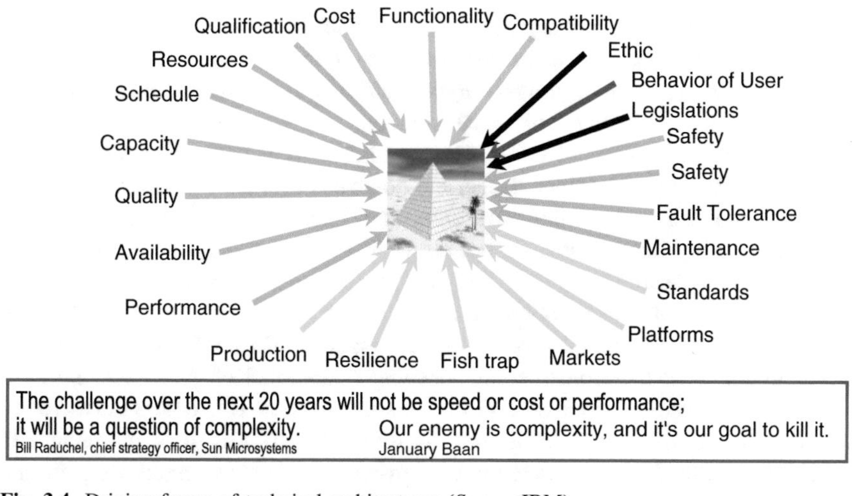

Fig. 3.4 Driving forces of technical architectures (*Source* IBM)

simple solutions and why some development processes are discontinued. It forces us to create highly sensitive cost-benefit calculations before we start to the product development process. Nowadays even single characteristics are analyzed by their value and we examine which characteristics are a basic need for which target group and which only an excitement factor. We could now look at examples for the driving forces but this would probably only lead us to the following conclusion: Depending on the market and the requirements of the driving architecture forces there will be product characteristics and features which represent a basic need for the acceptance of the product and others which cause so much excitement that they lead the product to success in many different ways. In reverse, each need that is not fulfilled can imply a risk.

This is why the architecture of product development is never developed solely based on safety requirements but also on other driving forces.

This means that the elements, from which a system is composed, are not defined only according to mere safety related aspects.

Of course, as already mentioned, money is an important factor. However, the availability of materials (e.g. rare earth material) production capacities, know-how, experience, transportation routes, supply chains etc. will also play an important role in how elements are defined and what position they will have within a system.

ISO 26262 only covers mere electric and electronic elements as well as software elements. However, a capacitor and a resistor alone will not be able to discover a low pass function. The elements of other technologies mentioned in the norm also play an important role. At least for the analysis of failure dependencies (Analysis of Dependent Failure or the better known aspect, the Common Cause Analysis) we see that connectors/plugs, printed circuit boards and housing can have a great influence on safety. Housing still represents a challenge for the project management in the automobile industry. In addition to the fact that they have to be ordered at the beginning of the project development process because it belongs to the so-called long-leading components (components which need to be ordered early in a development project, any change lead to an extraordinary extension of the development time) and thus defines the installation room for the control devices, but also because the arrangement of printed circuit boards and plugs need to then be defined prior. —What does this have to do with safety?

The metric of ISO 26262 only mentions random HW-faults. Some people say that the conductor paths, joints, cables and plugs can also have random HW-faults. However, we will see that this is not the main issue. Mostly it is the systematic failures, meaning the potential design errors that are most challenging. Plugs and conductor paths need to have a certain diameter in order to carry specific currents. The distances are also very important. Particularly for voltages beyond 60 volts; we will need to consider completely new safety aspects. ISO 26262 does not require de-rating (conservative construction, meaning operational characteristic are considerably below the nominal values of the components) as IEC 61508 does (70 %) but the characteristics need to be robustly dimensioned throughout the entire life time. This eventually determines the plug distances, pin size, conductor path distances, thickness etc. through the interpretation of safety mechanisms. In some cases this can cause a shortage of space in the housing. In reality this also leads us to the next issue—there is no current without dissipating heat (except from superconductivity). Each electric component creates heat, which has to be directed outside. The thermos conductivity of the housing plays an important role here. Overheating is a major cause for fire in control units. This is explicitly mentioned in ISO 26262 since this could be a failure function of the electronic.

The topic of heat plays another major role in a other typical long-leading component, the microcontroller. The higher we clock a microcontroller the hotter it gets. The amount of operations per time unit also influences the heating. This means a microcontroller that runs at its limits gets extremely warm. If we are able to redirect the heat we can push these limits but if other components are aligned too closely, we risk a heat buildup.

This shows that there are many factors that can influence the characteristics of a product. Besides the complexity of the dependencies from the example mentioned above, we see that other factors can also be extremely influential. Housing or a microcontroller does not get exchanged without a substantial reason during the development of a product.

Therefore, we have a great dependency on project management and architecture.

The book "Engineering a Safer World" by Nancy Leweson mentions a suggestion that describes the structure for various views of architecture and the allocation of requirements. All coherences are described with the term "Intent Specification" (Fig. 3.5).

The central statement is based on the idea that the product structure, the organizational structure that the product should create as well as the management structure need to be well matched. In order for the respective organizational units to

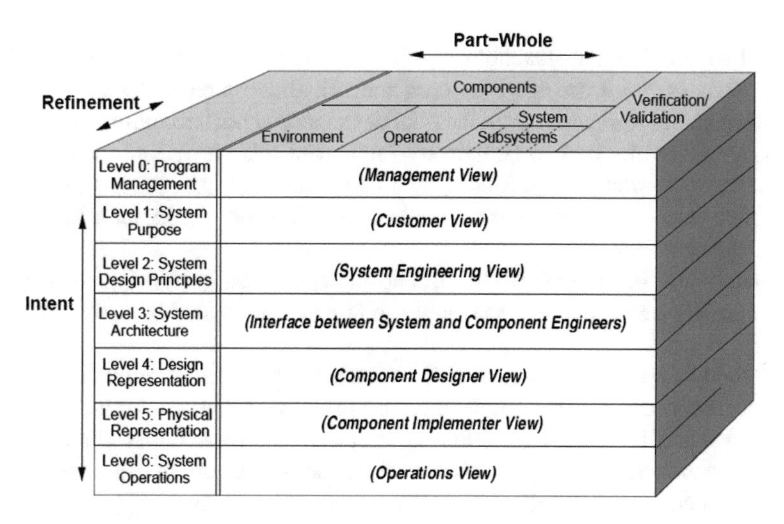

Fig. 3.5 Multi-dimensional structure of specification (*Source* Figure 10.1, Nancy G. Leweson [3] Engineering a Safer World)

work with each other it is necessary that this structure also representative of the foundation for the specification.

Therefore, product architecture becomes the most important element for the foundation of the product structure.

As a consequence, the first step of project planning is to create a project structure tree, which considers the following aspects:

- Product, organization and project interfaces need to be well-matched. The more interfaces there are for the three interface dimensions, the more complex the product development.
- Product, organization and project interfaces need to be defined and controlled through a hierarchical arrangement. Each interface needs to be defined and managed in an higher hierarchical level.
- The product structure and the horizontal and vertical interfaces create the basis for the specification of the elements of the architecture and their behavior or the dependencies among them.

The cube structure of Nancy G. Leveson is no longer discussed in this book. Only the product technical viewpoint is further considered. The organizational structure or the viewpoint of customers or suppliers—those are considered in the product planning, project planning and determination of organizational interfaces. However, these aspects form the basis for the planning of safety activities in the project safety plan as a derivation of the safety lifecycle (compare to ISO 26262, part 2, chapter 6.4.3, “Planning and coordination of the safety activities”).

3.3.2 Views of Architecture

Since there are different stakeholders of architecture there also need to be different abstractions of the description for each individual stakeholder. Ideally, we could abstract certain information from the total description model according to the profile of stakeholders. In order to completely implement this we would need to have a standardized version of stakeholders and their often varying interests and besides that also a basis data model, which would be able to conceptually include the entire world and its coherences.

Nobody has time to wait for this to happen. Even such genius data management and information systems such as Google, Wikipedia etc. would be stretched to their limits.

Anybody who has ever tried to build a house knows a construction drawing. The aim of this construction drawing is of course to show the later owner how the house will look once it is finished. Often, particularly for houses that are built by a building contractor, there are also construction specifications, but the average person is unable to read this without the help of an attorney.

The construction drawing often shows a front, back and various side views and vertical sections in order to see the arrangement of the different stories as well as horizontal sections to see the arrangement of doors, for example. However, the drawing always shows us the same house. Our expectation is that the different views are consistent and that we can see the entrance door in the front view on the same place in the house as we would expect to see it from the horizontal view.

Nevertheless, different stakeholders of architectural company need to be identified and all of them of course only want to see the view of the architecture that interests them. Therefore, if we send the carpenter the plan for the inside doors he will be interested in knowing the height of the floor fill but not the allocation of door lintels and how much iron was used for which ultimate load. Actually, here we would need to consider the perspective of the finance controller and the project manager since the resources used already determine the safety sufficiency.

Towards the end of the '60s, Phillipe Kruchten described his four views, which lead to the following (here compared to UML) 4 + 1 views:

- The logical view describes the functionality of a system for the end user. Logical elements are used in order to show different dependencies of elements. Class diagrams, communication diagrams and sequence diagrams can be used as UML-diagram.
- The development view or implementation view describes the system from the viewpoint of the developer. Component diagrams or package diagrams can be used as UML-diagrams.
- The process view (behavior or functional view) describes the dynamic aspect of systems as well as the behavior of elements at their intersections to each other and in a defined environment. Relations could be any kind of communication (technical but also man-machine communication etc.) time behavior as well as allocation and structure aspects such as parallelism, distribution, integration, performance and scalability. Activity, sequence or timing diagrams can be used as UML-diagrams.
- The physical view or the deployment view describes the system from the point of view of the deployment or rather the manager of deployment. It should include the allocation of components, modules or electric components and elements that have to be obtained and deployed for the communication among each other (as for example cables, bus, plugs…). Distribution diagrams can be used as UML-diagrams
- The scenario view describes the planned cases of application, possible configurations and behavior versions. This can be the basis for the planned behavior of elements among each other. The architecture verification later forms the foundation for integration tests. Use-Case diagrams can be used as UML-diagrams.

In the context of the funding project "Safe", views and perspectives for the automobile industry were derived from definitions of the project SPES2020 (see Fig. 3.6).

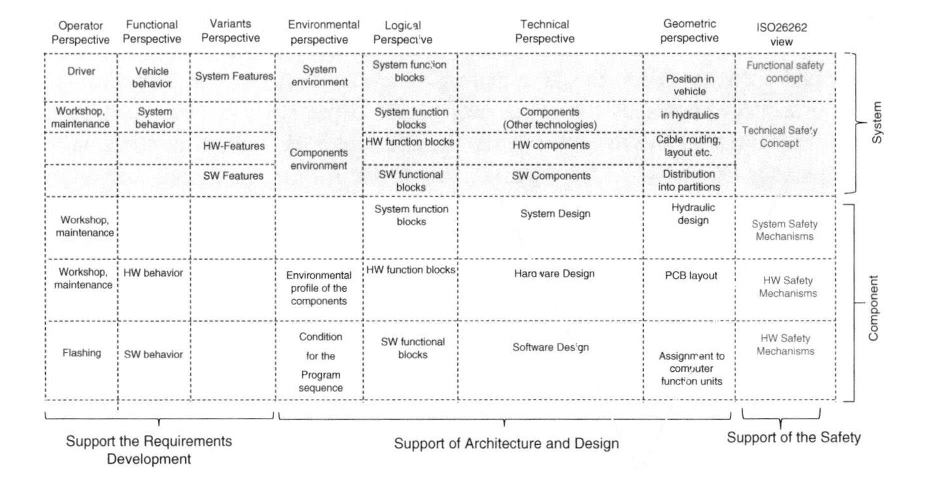

Fig. 3.6 Perspectives of architecture (*Source* Funding project "Safe")

The individual perspectives can be described as follows:

- The operator perspective describes the behavior interfaces between humans and technical systems and their elements.
- The functional perspective represents the observable technical behavior.
- The version perspective describes the dependencies or differences of various characteristics or implementations from the viewpoint of the respective stakeholder of the system or its elements. In this case a stakeholder can also be a system or an element.
- The logical perspective uses logical elements to illustrate interfaces or behavior on such interfaces.
- The technical perspective uses technical elements to illustrate structures and interfaces or behavior on such interfaces.
- The geometric perspective shows the position of a system or its elements in a certain context or environment.
- The safety perspective shows the safety relevant aspects of an architecture.

3.3.3 *Horizontal Level of Abstraction*

Abstraction is often paraphrased as the omission of individual component and the transfer to something more general and simple. Depth is considered to be the horizontal level of abstraction in which we practically look into a car. The idiom "To miss the forest for the trees" can be used as an appropriate description of the challenges for the development of vehicle functions.

When describing the behavior of a vehicle there are dependencies that can be broken down all the way into the individual lines of a software code, the resistor or joints of components on the printed circuit board. If those dependencies are not recognized we have to ask what other elements are needed.

Consequently, in vehicle and aircraft engineering we speak about the airplane level, the system level and the components level. If this were a developed requirement for the structure the development of vehicle functions would most definitely be a lot easier. Officially, these levels are not mentioned in ISO 26262 but the norm helps to better understand when would be a good time to take such levels in considerations (Fig. 3.7).

Here we can see that the environment for a vehicle and an airplane has an essential influence on the development of a system. The degrees of variance for the vehicle alone are a lot less for the steering system than for an airplane. However, even for a vehicle the degrees of variance are very different. A motorcycle will fall easily while a car won't, unless we refer to an elk test. This comparison shows that certain events have a design related probability of occurrence. On the system level, which shows the interaction of the components, also mechanic, electronic and software components are applied but based on different requirements and environmental parameters they will lead to extremely different system designs.

If we consider the components level we will find a similar situation for electric hardware and software. In this case the differences will mainly show in the architecture.

Architectural patterns show that the functions in airplanes, approach automobile architecture more and more frequently. So far comparing systems and voter (for example a 2 of 3 selective system) were mainly deployed on the system level and

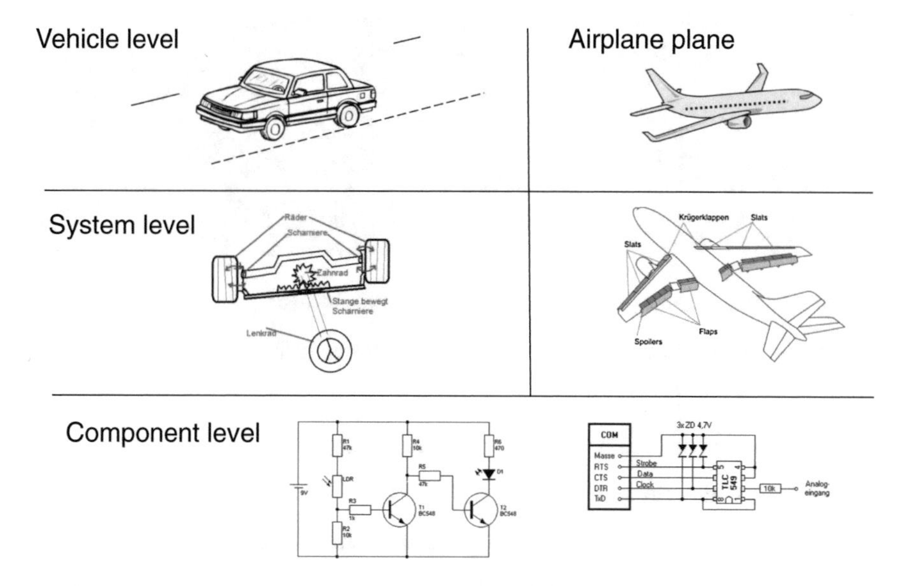

Fig. 3.7 Comparison analogy vehicle level/airplane level to system/components level

implied by for example, three independent components (also called device redundancy) which then implement safety functions through an independent majority voter system or passive logic elements like relays, diodes, switches etc.

For approximately 20 years the automobile industry knows the EGAS-concept. (E-GAS or E-Throttle) It implements redundant software levels, which then prioritize safety functions as needed, send signals through enabled pathways to a safe state or an intelligent watchdog switches-off the entire computer. In regard to aircraft manufacturing we refer to a command-monitoring-system. The commonality of these concepts is that the target functionality should function independently from the monitoring function. The design aim of such a monitoring function is to realize it in a way that in case of its failure no harm occurs through the product. Such principles of redundancies have developed over time and are described in ISO 26262 as ASIL-decomposition. Here these comparisons or voting can be deployed as completely independent or sufficiently freedom from interference software functions. In order to achieve such independencies and/or absence of interference from system or hardware measures are still necessary but the aim is to avoid components or control device redundancies. Semiconductor manufacturers that provide respective built-in-self-test (BIST), diagnosis, memory partitioning or redundancies (diverse I/O-periphery or multiple-core-devices) on a basis chip, support this development. The comparison with the aircraft industry shows no clear boundary and on what parameters do we need to allocate the horizontal cuts within the architecture.

Regarding system integration ISO 26262 mentions three (horizontal) integration levels. In part 4, chapter 8 the following targets are determined:

ISO 26262, Part 4, Clause 8.1.1:

8.1.1 The integration and testing phase comprises three phases and two primary goals as described below: The first phase is the integration of the hardware and software of each element that the item comprises. The second phase is the integration of the elements that comprise an item to form a complete system. The third phase is the integration of the item with other systems within a vehicle and with the vehicle itself.

This results leads to three horizontal integration levels:

- Integration of the vehicle system (items) in the vehicle
  - vehicle interfaces
- Integration of the components to a defined system
  - component interface
- Integration of electronic hardware and embedded software
  - hardware-software interface (HSI)

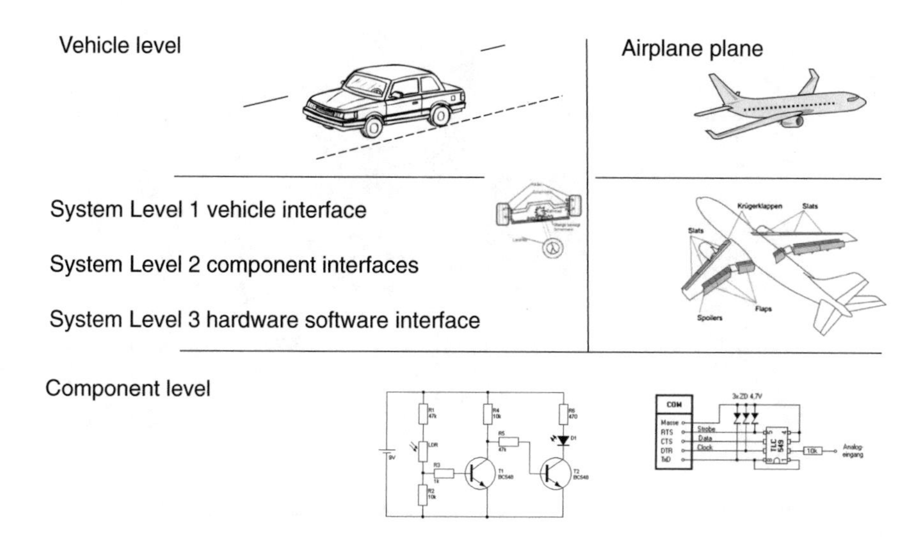

Fig. 3.8 Multiple system level between component and vehicle level

Since the interface to these levels of course have an influence on the architecture and the necessary requirements for these levels, these levels also need to be considered in the system requirements development. This means the architecture needs to be planned accordingly so that these horizontal interfaces already exist. Figure 3.8 shows these three levels embedded between the component level and the vehicle level.

System level 1 orientates itself on the interface of an item (vehicle system). In this case many decisions and definitions are already made that can have an essential influence on the later component deployment. All requirement covered in ISO 26262 part 3, chapter 4 "Item Definition" can be important for the components.

According to the Ford-FMEA-handbook [5] there are four kinds of interfaces. Here they are shown and described with examples (Fig. 3.9).

The following factors can be seen:

Physical interface

- geometric data, which describes the space in the vehicle in which the components need to be integrated
- environmental conditions such as vibrations, temperature, dirt
- physical values or limitations such as force, torque, turn rate, positioning angle, transmission ratio, and their tolerances
- electric values such as voltages, currents, EMC, data interfaces
- kind or type of data (physical, electrical etc.)
- data formats, data contents, signal level
- data intersection, bus or communication systems (CAN, Flexray, Ethernet)
- Network or bus topology (star, ring, node, gateway)

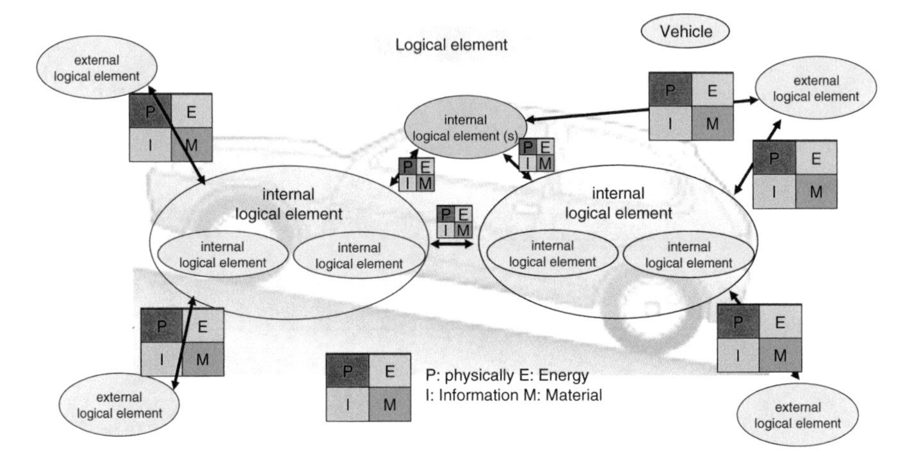

Fig. 3.9 Multi-dimensional boundary and interface analysis (*Source* Derived from Ford-FMEA-Handbook)

Energy interfaces

- type/kind of energy, such as electric, kinetic energy or pressure or vacuum
- energy transfer such as voltage levels, short circuit currents, safety deployment
- energy amounts such as capacity of batteries or capacitors
- kind of energy provision such as via cable, induction or

Material transfer (interface)

- fuel delivery, lubricants etc.
- material compatibility such as hard/soft materials, type of oil, chemical compatibility (salt, sulfur with iron etc.)
- mass shifting, loading conditions

These interfaces can also show time dependencies. It is important to lead the information to the brakes that the vehicle should stop but it also has to be ensured that the actuator is provided with sufficient energy. For the hydraulic brakes this is mainly ensured by the brake pressure. For example electric power supply can, at a specific load, no longer provide sufficient energy or the fuse trips at a certain threshold.

At system level 2 descriptions of the intersections can vary, also the time requirements often become more detailed and thus mostly shorter. An example in steering shows this hierarchical cascading. A steering system is able to tolerate an error at a certain pulse width and energy for about 20 ms. If the impulse lasts longer the driver won't be able to control the vehicle any longer and might drive into the oncoming traffic. This means the safety tolerance for such a system is approximately 20 ms. If we break this down to the control unit this time can be reduced to below 5 ms. So from connector pin to connector pin, the control unit needs to initiate a safety related correct reaction within below 5 ms. In order to ensure the same even at system level 3 for the hardware-software interface a microcontroller needs to initiate a software function below a millisecond, which can present an adequate reaction at the pin of the microcontroller.

On system level 2 these interfaces can be described as follows:

Physical Interfaces

- geometric data in the housing such as the attachment of plugs, printed circuit boards…
- environmental conditions such as vibration, temperature, dirt (this data can vary since sensors, control units or actuators can be implemented in different places or be protected by the housing of pollution or humidity, reduce vibrations, dissipating heat)
- physical values or limitations such as force, torque, turn-rate, transmission ratio and their tolerances (these values can again be broken down or allocated to different elements of the system)
- electric values such as voltages, currents, EMC, data interfaces (see above—physical values)

Information interfaces

- type of information (information here often more specified)
- data formats, data contents, signal level (itemization of information)
- data intersection, bus or communication systems (CAN, Flexray, Ethernet), now the physical specification of the communication interfaces is required so that internal and external communication partners can communicate with each other
- Network or bus topology (Star, Ring, Node, Gateway), here these elements are specified in detail

Energy interface/intersection

- type of energy, such as electric, kinetic energy or pressure or vacuum
- energy transfer such as voltage levels, short circuit currents, safety design characteristics deployment
- energy amounts such as capacity of batteries, capacitors etc.
- type of energy provision such as via cable or induction

These interfaces are now broken down into the individual external and internal components and detailed as needed.

Material transfer (interfaces)

- fuel delivery, lubricants
- material compatibility such as hard/soft materials, transmission or hydraulic type of oil, chemical compatibility (salt, sulfur with iron etc.)
- mass shifting, loading conditions

These interfaces are now as well broken down into the individual external and internal components and detailed as needed

Often, in system level 3, new information appears which derive from the specification of the microcontroller. But again, also here, all four interface-categories are either more or less relevant. Material transfer will be less relevant for the microcontroller and more important for the material interfaces. Here we can find contacting issues because of wrong materials up to drift or sporadically effects due to corrosion.

Components Level

To define mere mechanic components in various abstraction levels could be in some cases reasonable but generally, complex mechanic components are already described on system level. Also the interfaces can be allocated to any given system interface. A sheer hydraulic steering will therefore more likely be integrated in system level 1 and parts such as printed circuit boards, plugs etc. probably in the components level.

In the case of software it often occurs that multiple software components exist, which are then integrated into the entire embedded software in a microcontroller. Formally seen, multiple functional groups that are integrated into a microcontroller can be integrated as system elements. However, because the hardware-software interface imposes a lot of requirements on the software that need to be implemented, the interface is becoming increasingly complex (Fig. 3.10).

Time analyses for such integrations can only be analyzed through alternative views since the runtime environment, time-management or partitioning, clock-frequency etc. of the computer need to be considered for each software

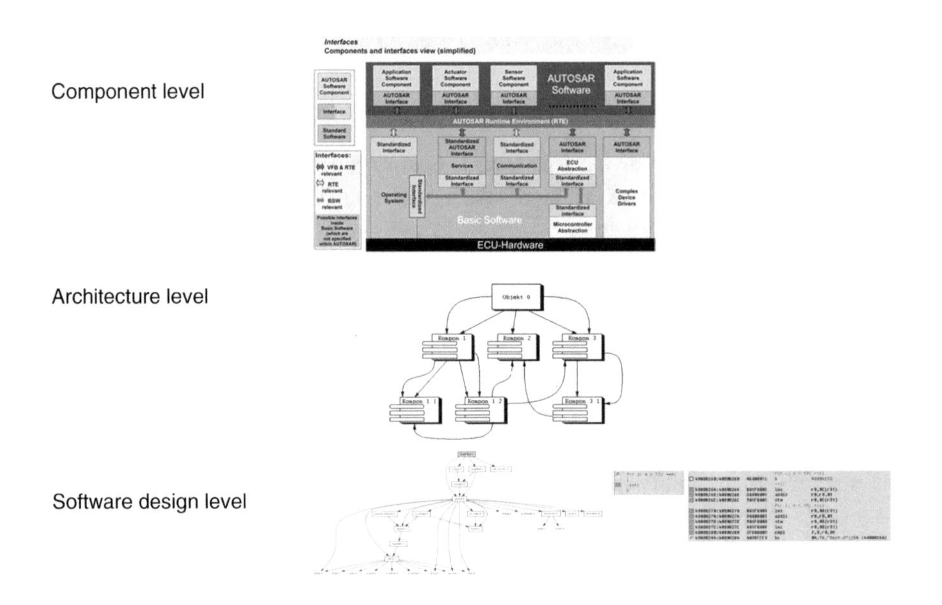

Fig. 3.10 Software architectural layers similar to Autosar

element. Here we have the so-called 'software architecture level', which forms the level between the software design and the software component. Just like in SPICE the term software unit is seen as the smallest entity of software. Therefore, instructions are no longer considered to be an entity. In the C-programming language, which is often used in the automobile industry, the C-file would represent such a level of abstraction for a software unit. Furthermore, in the software development two different levels are differentiated the basis software and the application software. Interfaces that provide a defined data structure during runtime, a so-called runtime environment (RTE), as we understood from AutoSar, show a separation between basic and application software.

However, in the electronic field we still mainly start with mechanics. First of all we have the housing, plugs, connectors, printed circuit boards, air ventilators and cooling devices. These already specify some parameters and design restrictions for the deployment. Since the housing also has to be ordered at an early stage of the project development, the entire electronic deployment will depend on it. Formally this is considered to be the abstraction level of the electronic architecture. Therefore those design dependencies should be known but not specified as an independent abstraction level. However, these mechanical components can be reasonable separations for various electronic components: several electronic components on different printed circuit boards, separation of control and power electronics, different voltage levels or also technical separations of safety-relevant electronic and non-safety-relevant electronic.

For software however we reasonably separate, if varying software components are integrated in different microcontrollers. For electronics we also often need to find other ways to separate components, functional groups and components. This is why in electronics there are three abstraction levels known as components level, functional group level and components level. Semiconductors such as microcontrollers, ASICs, FPGAs or other hybrids are often integrated as functional groups even if they count as components (Fig. 3.11).

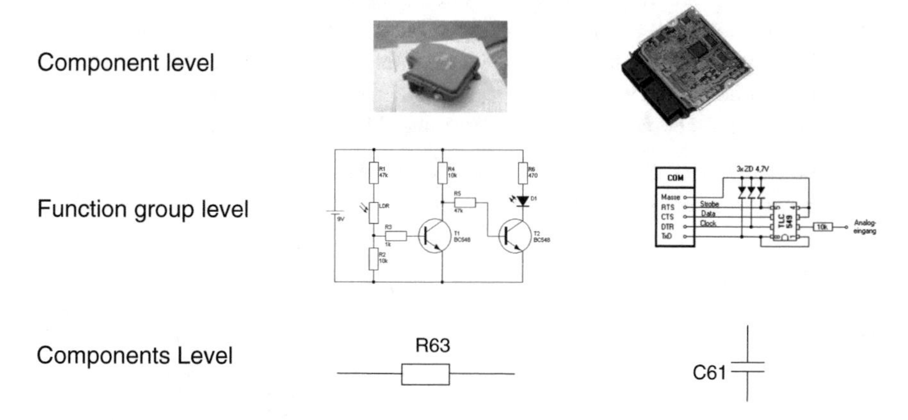

Fig. 3.11 Horizontal layer of abstraction in electronic hardware

Software: Basic Software—Application Software

Also for basic software or in AutoSar abstraction levels are addressed but in this case we don't mean horizontal abstraction levels but functional (perspectival) abstraction levels (e.g. hardware abstraction or microcontroller abstraction layer (MCAL)). Nevertheless, the interface between application software and microcontroller still play an important role in the definition of the abstraction level. In early revisions of ISO 26262 the hardware software interface (HSI) was implemented in part 5 (Product development at the hardware level) and 6 (Product development at the software level) and only after the CD version of ISO 26262 it moved into part 4 (Product development at the system level). What is special about this is that the microcontroller as a hardware element, similar to the electronics housing, predetermines essential design characteristics for the software. In order for these two components to interact properly, those characteristics and their potential flaws as well as the functions and their potential failure functions need to be considered. This obviously goes for all component interfaces. In the case of HSI we find a lot of relevant interface parameters. This means, that it is not only about the correct functions of the so-called low-level drivers, which provide information on microcontrollers to the software, the operating system, peripheral (DMA, I/O, bus etc.), internal communication, logic unit, memory or function libraries, which are provided by the computer, but also the systematic protection of potential failure or malfunction at this interface.

3.4 Requirements and Architecture Development

The architecture should also display the structure of requirements. With the help of the horizontal abstraction levels described the upper and lower interfaces are predetermined for the details, which the architecture should illustrate. By determining the logical and technical elements, further interfaces within the horizontal abstraction level are displayed. In a system, logical and technical elements are defined, which have the task of carrying or also implementing the required functions. The logical or technical elements have to be clearly specified so that a correct behavior can be expected for safety relevant systems (Fig. 3.12).

The following characteristic or features should be seen as requirements:

- The environment, in which the element should be embedded, has to be specified in a way that all influential factors, which can influence the behavior of the

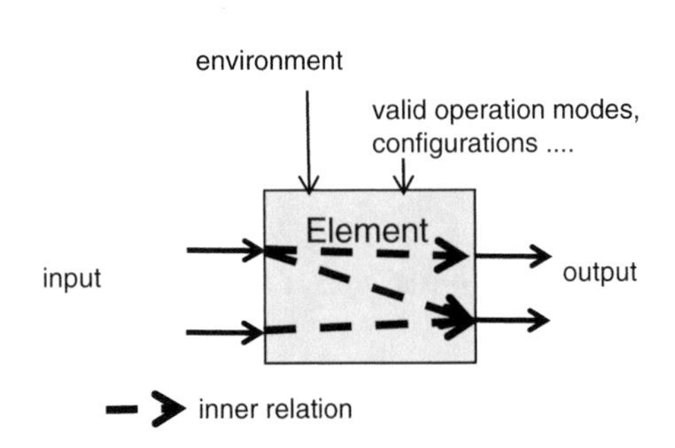

Fig. 3.12 Types of specification items for complete element specification

element, are defined. What factors need to be considered are the result of an interface analysis (in terms of a boundary analysis).
- The permitted ways of use, modes of operation (as for example initialization, monitoring, on-demand, stand-by, regular operation) or configurations (e.g. only for analog data processing, calls with certain parameters (e.g. for software components), clocked or triggered processes or functions) have to be specified as well as the kind of way how information are allocated to the element.
- The input information should be specified so that it can be determined where they are generated in which format they are transmitted and in what range the information is valid.
- The output information should be specified in a way that it is defined where they are to be addressed, in which format they have to be provided and in what range the information is valid.
- The internal relations should define all input and output conditions in the specified environmental conditions under the permitted mode of operation or configuration. If memory effects in the elements can change the internal relations, they also have to be defined through the specification. Memory effects within the elements leads to changes in the input and output relations, which have to be defined.

Besides the functional characteristics the following characteristics will have an influence on the electric, electronic and mechanic hardware elements:

- Geometric, form, volume, mass, structure, surface, labeling, color etc.
- Material properties (material compatibility, chemical reactivity)
- Behavior and reaction towards physical influences such as temperature, electricity, voltages, stress behavior (vibration, EMC, certain behaviors referring physical stress)
- Aging effects (statistical aging behavior (Weibull-binominal-, chi-distribution))
- Maintenance requirements, logistic
- Time aspects

But also technical software elements have technical characteristics such as

- Size of the compiled code
- Branches, memory consumption, quantity (number of instructions, variables, addresses, jumps, calls, interrupts)
- Realized program flow, task allocation, scheduler strategy, etc.

Commercial, ideational or emotional aspects are not further regarded here since they should not consider to be related to safety. How these technical values are to be defined and specified is generally also the result of an analysis.

We can assume that a technical element has specific characteristics; otherwise it couldn't fulfill the intended characteristic or function. Therefore, it is only logical that if those characteristics are no longer given or consistent, functional limits arise (Fig. 3.13).

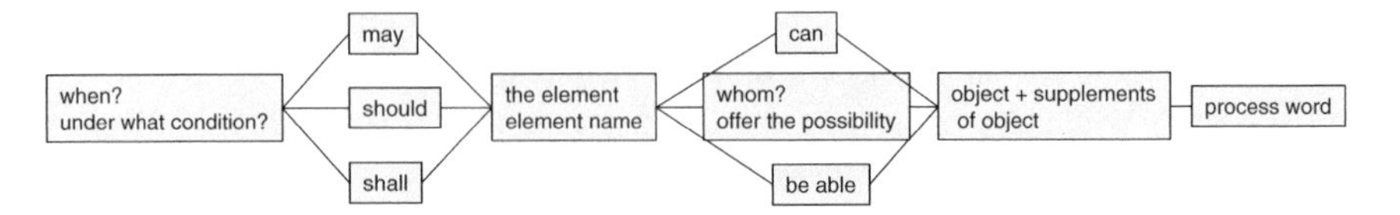

Fig. 3.13 Requirement template (*Source* Based on Chris Rupp [4], Requirement Engineering and Management)

Templates should be designed for all these requirement aspects so that requirements are displayed in a clear format. This avoids wrong interpretations and ensures the consistencies with the architecture. Specification in natural language doesn't mean that all lists of characteristics need to be formulated in verbal sentences. Semi-formal methods are better suited particularly for technical behavior and are clearer and less unmistakable than well-formulated sentences. In a well-structured architecture a lot of requirements are automatically derived from the architecture because of such templates. Or the essential content can be formulated in advance through the definition of keywords so that only parameters or certain characteristics need to be added. All signal flows or data flow aspects need to be consistent with the architecture, meaning that if these requirements are automatically derived from the architecture, a good level of consistency can be expected.

3.5 Requirements and Design Specification

All requirements for "Requirement Management" refer to the specification of requirements in ISO 26262, Part 8, clause 6 (Specification and Management of Safety Requirements). Requirements for Design Specifications could hardly be found in norms and standards at all. Most likely, the only requirement on Design Specifications in this case is that the content needs to be understood. The challenge is to find a healthy mix between requirements and design and specify it correctly and sufficiently (Fig. 3.14).

The example with the image of the Mona Lisa shows that a mere statement of requirement can become pretty extensive. The recipient of this statement of requirement will have a hard time creating this image with the given information. A good mix of requirements and clearly stated design characteristics, which are also illustrated respectively, can be helpful.

For a mechanic design it would not be suggested to specify a M6 screw with requirements specifications. Furthermore, one would not voluntarily write requirements specifications for a resistance with 100 Ω and a tolerance of 1 %. Now the question is if this is so obvious for electronics and mechanics, how do we define the limit for a system or software? The example of Mona Lisa shows clearly that requirements specifications alone are not enough. Now how do we structure specifications and to whom do we assign them? The specification of a vehicle

Fig. 3.14 Requirement specification and design specification

system is generally not assigned to the driver but should be directed to the technician. This means for a system developer timing diagrams, spreadsheets, sequence diagrams etc. should be significant information. To specify these requirements one more time in natural language is inefficient and unnecessary. The clear description of technical behavior is often also easier to explain with models. Basically, the picture of the Mona Lisa is nothing else but a model, which completes the requirements. The system design chapter of part 4, chapter 7 requires a system design specification for systems and for the software in part 6, chapter 8, a software design specification but no requirements specifications.

ISO 26262, Part 8, chapter 6.2.1–6.2.4:

6.2.1 Safety requirements constitute all requirements aimed at achieving and ensuring the required ASILs.
6.2.2 During the safety lifecycle, safety requirements are specified and detailed in a hierarchical structure. The structure and dependencies of safety requirements used in ISO 26262 are illustrated in Figure 2. The safety requirements are allocated or distributed among the elements.
6.2.3 The management of safety requirements includes managing requirements, obtaining agreement on the requirements, obtaining commitments from those implementing the requirements, and maintaining traceability.
6.2.4 In order to support the management of safety requirements, the use of suitable requirements management tools is recommended.

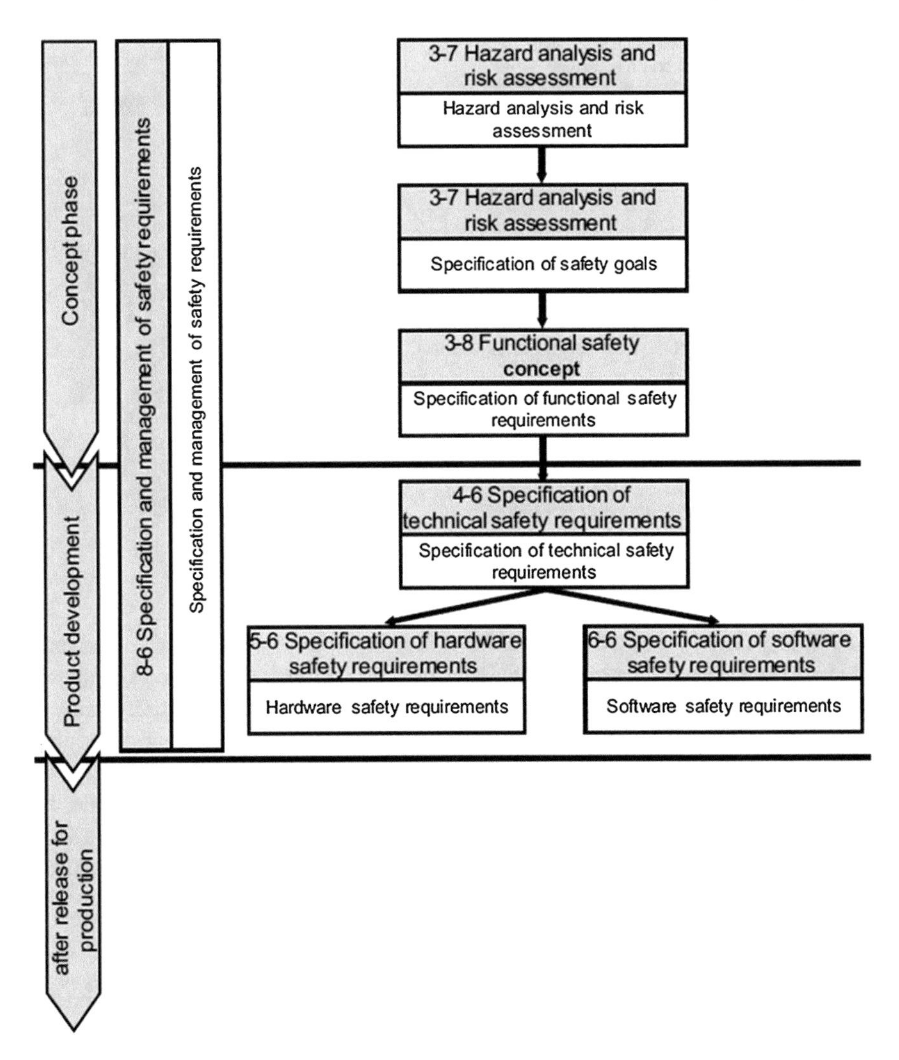

Fig. 3.15 Structure of requirements (*Source* ISO 26262, part 8, Fig. 2)

The following Fig. 3.15 is an excerpt of the safety lifecycle and shows how activities, requirements and work results leave the conception phase and enter the development phase.

ISO 26262, Part 8, Chapter 6:

6.4.2.3 Safety requirements shall be allocated to an item or an element

The following Fig. 3.16 clarifies the requirements for requirements and the way engineers (requirement engineering) should manage them.

There are now requirements in the common safety or development standards that also characteristics of design elements need to be specified as requirements. However, in all architecture chapters requirements play an important and central

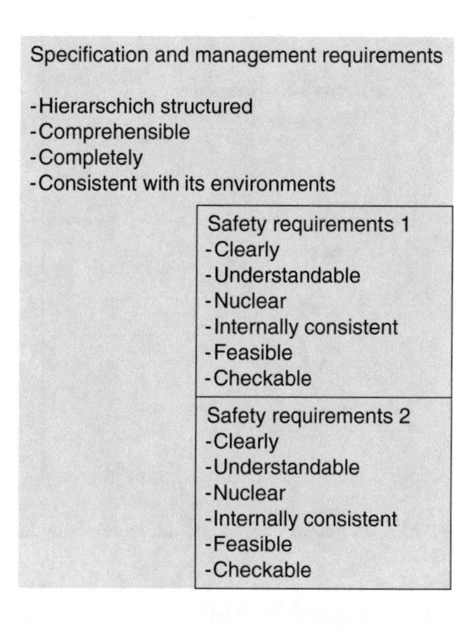

Fig. 3.16 Relationship between management of safety requirements and requirements (*Source* Based on ISO 26262, part 8, Fig. 3)

role. But if a software unit and an electronic component do not need to be completely specified with requirements, how can we assure everything will be complete? The target is to find a level that defines all behavior sufficiently and unique. This level needs to be planned and clearly described in a requirement and architecture strategy (requirement and architecture strategies are also work-products addressed within the base practices from SPICE).

Technical products or parts of it such as their characteristics, limitations, constraints, range of use, area of application, behavior etc. should always be specified with a reasonable mix of requirements, architecture and design requirements. Products are never described by their errors or risks. Nevertheless, it can be necessary to document them for the end-user (package information, manuals etc.).

Functional Architecture and Verification

A function is generally seen a mathematical/arithmetic expression or relation/coherence.

$$f(x) := ay + bx$$

This is a typical mathematic function. For systemic functions there are the following illustrations for the following functions: (Fig. 3.17).

All 3 illustrations represent the same relation; however, there are only different ways to present the information that function 1 is composed of 3 partial functions. Function 1 represents the sequential chain of 3 partial functions. These sheer functional perspectives allow no identification of interfaces of the system or element boundary. Only a simple and limited description of the behavior of technical

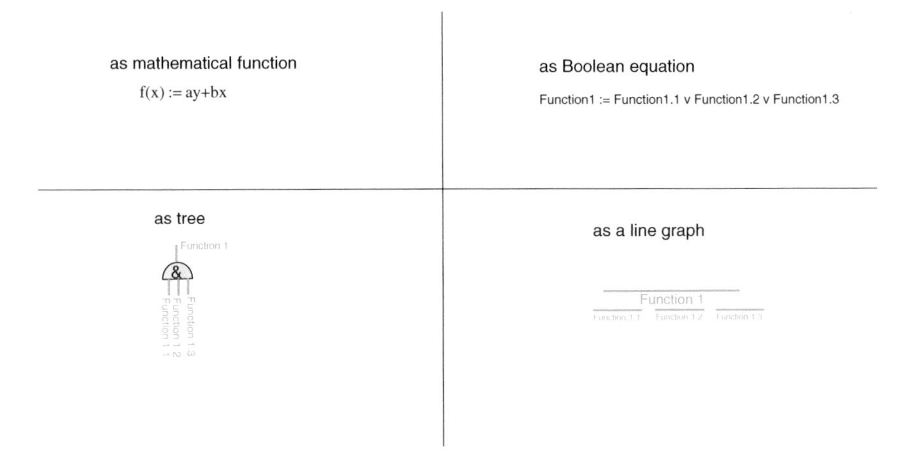

Fig. 3.17 Function decomposition as mathematical or Boolean function, tree or line diagram

systems is possible. Also the mathematical transfer function describes an expected result based on defined inputs (Fig. 3.18).

The mathematical transfer function already considers inputs and outputs. Therefore, interfaces are already available here. Also in modeling tools like Matlab-Simulink-Model the input and output relations are foundations for the interfaces of the architecture.

If requirements of one element are derived from an inner structure, new requirements for interfaces emerge (Fig. 3.19).

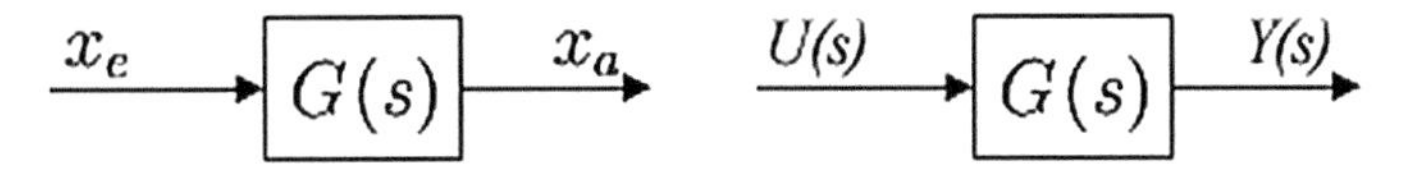

Fig. 3.18 Mathematical transfer function

Fig. 3.19 Deriving requirements and allocation to functions or on logical, functional, or technical elements

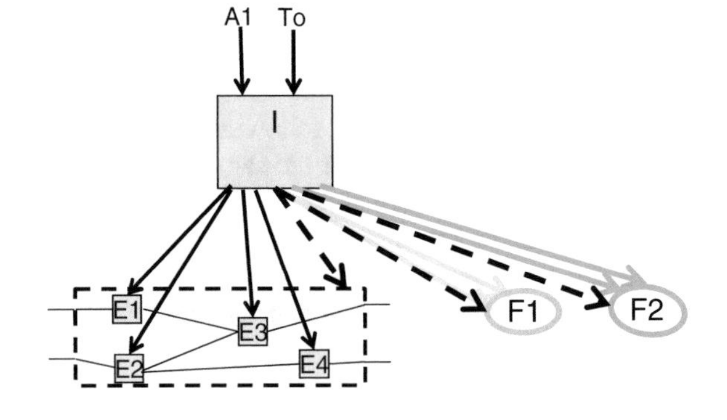

Allocations of functions, partial functions and their requirements (so called functional requirements) on a logical element are the main activity for the development of the functional safety concept besides the verification of such requirements. Without such allocations verification is impossible. The logical elements E1 to E4 should implement function 1 and 2. The allocation could lead to the following result:

Logical elements have limitations and identifiable interfaces but also the functions obtain interfaces and limitations through the allocation of the logical elements.

In this structure and with the information and correlations given, requirements can be verified. The following verifications of requirements would be possible:

Have all requirements been derived?
Were the requirements classified in a way that it is clear whether they are requirements for an input signal, output signal, a relationship within an element, a relationship between two elements, a function between two elements, the environment of elements or design requirements or limitations?

Similar approach as in Fig. 3.20, requirements could be derived only from a higher element or are there other higher elements outside of the system boundaries that can influence the requirements?

Is the internal structure of the derived elements described sufficiently?
These questions build the foundation for the verification of the functional safety concept. In each individual level, in which requirements can be verified, a similar approach can be used for the requirement verification. The figure above shows that if functions or the elements, which those functions should realize, do not have common interfaces, the number of interfaces will explode exponentially. If also a situation related failure analysis had to be made on the basis of such heterogenic positive descriptions and perhaps correlations had to be described through several horizontal abstraction levels, completeness, transparently, comprehensibility, consistence and correctness would no longer be given. Such an amount of interfaces would not be analyzable and therefore no longer controllable.

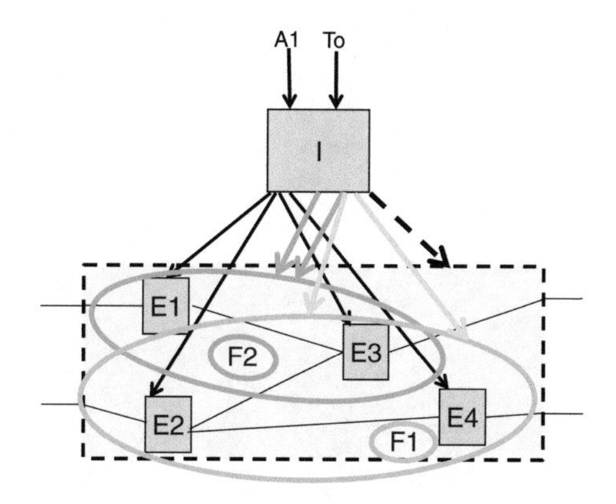

Fig. 3.20 Allocation of functional requirements, functional elements or function groups on elements

References

1. [DIN EN 61709]. Electric components - Reliability - Reference conditions for failure rates and stress models for conversion (IEC 61709:2011); German version EN 61709:2011
2. [SN 29500]. 1999, Siemens Standard SN 29500: *Ausfallraten Bauelemente*
3. [Nancy G. Leweson]. N.G. Leveson, Engineering A Safer World: Systems Thinking Applied to Safety, MIT Press, Cambridge, MA, 2011.
4. [Chris Rupp]. Chris Rupp & Die SOPHISTen: Requirements-Engineering und -Management: Professionelle, iterative Anforderungsanalyse für die Praxis. 5. Auflage. Hanser, 2009
5. [Ford-FMEA]. Ford-FMEA-Handbook, Ford Motor Company, 2008
6. [ISO 26262]. Road vehicles – Functional safety. International Organization for Standardization, Geneva, Switzerland.

Chapter 4
System Engineering for Development of Requirements and Architecture

The ascending branch of the V-model has not always been intensively and systematically implemented in the development process of vehicle components. Crucial indicators for the automobile industry are methods such as statistical design of experiments (DoE) or an intensive validation. The descending area of the V-model has often been neglected. Writing specifications is not strength of automobile manufacturers.

As we have previously seen in the architectural views and abstraction levels of architecture, horizontal and vertical interfaces and also other differing views are structuring criteria. This applies particularly to the development of requirements. If we determine functional, technical and logical elements we also need to describe and specify them. If such elements are combined in order to function together as desired and create an intended function, they have to show compatible interfaces that are specified sufficiently. ISO 26262 [1] covers the "specification of interfaces" but does not clearly illustrate the respective requirements. However, the correlation between the work results such as requirement specifications in part 10, are shown based on information flows. In this case only the general abstraction levels system and components are covered and also the differing views on how a system can be described are not covered.

Figures 4.1 and 4.2 are published in DIS (Draft International Standard, previous version of the norm) of ISO 26262 part 10 (Figs. 7 and 8). Figure 4.1 was designed for the electronic hardware. It shows the horizontal and vertical interactions and information flows. In the requirements and design phase the direction of the arrow for the vertical axis points from top to bottom. In the integration and test phase it is the other way round. This illustration does not include any iteration in the safety lifecycle, which can become necessary because of model phases, change requirements or verification or validation measures.

For the software development one more level is illustrated (Fig. 4.2). It shows an architecture level and a level, which is assigned to the software unit or the design of the software unit.

H.-L. Ross, *Functional Safety for Road Vehicles*,
DOI 10.1007/978-3-319-33361-8_4

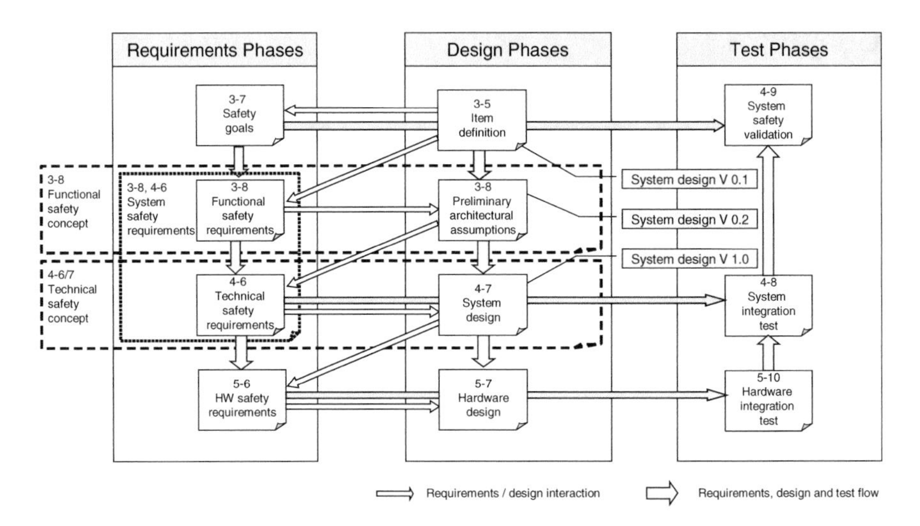

Fig. 4.1 Data flow in system and hardware product development (*Source* ISO DIS 26262)

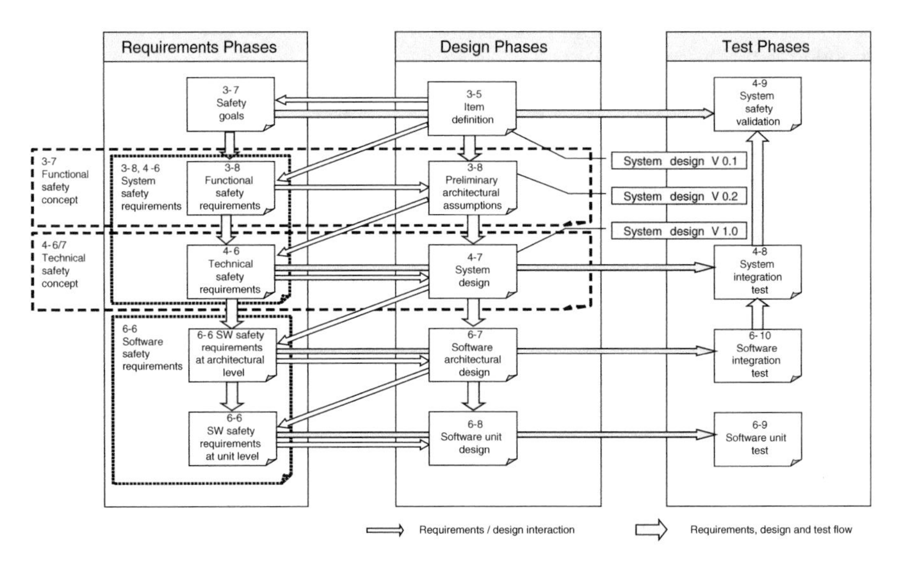

Fig. 4.2 Data flow in system and software product development (*Source* ISO DIS 26262)

In the final version the arrows and the keys are changed according to Fig. 4.3. The essential input for the definition of the vehicle system (ITEM) is the functional concept. Since ISO 26262 does not cover the hazards, which result from a correct functioning system, the functional concept itself is required to completely describe the free from danger of intended function and its structure. This won't be possible at an early stage of the product development. Therefore, we are forced to see all activities as continuous iterations for which the obtained insights have to be proven

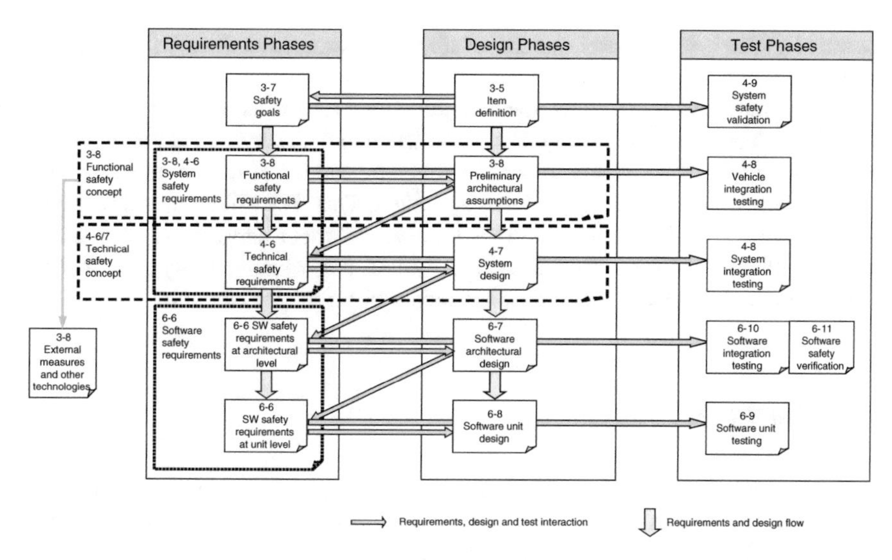

Fig. 4.3 Data flow in system and software product development (*Source* ISO 26262, Part 10)

based on the present output and results again and again. For the horizontal abstraction level such as the vehicle level, system level, components level, structure in silicon or other complex components or elements a similar, a deductive development process should be chosen in order to develop detail structures and the respective requirements. All levels contain logical, technical and/or functional elements, which interact with each other. Functions develop through an—according to specifications—correct interaction of elements. Those functions can often then are inductively verified, meaning back up the descending V-model branch. If in the process of the verification, safety requirements, which are already verified in the upper level, are confirmed and proven to be fulfilled, the verification serves as a sufficient argument for functional safety. This means, that on any given level the same principles of safety engineering can be applied and a higher consistency of development work can be achieved. The fact that on each level the sufficient control of errors is essential for functional safety does not mean that all errors in each level need to be controlled separately.

Since ASIL C, ISO 26262 also requires the control of multiple-point failures. For complex systems with various independent safety goals a systematic failure control can no longer be implemented. Therefore, safety goals which require two safety mechanisms need to be implemented in different horizontal levels. Through the implementation of barriers single point faults become multiple-point failures. Therefore, a single point fault in a sub system is often only a multiple-point failure in the overall context if failure propagation could be avoided to a higher level above the sub system. If multiple system elements fail at the same time, multiple-point failures occur. Also in this context the classification of failure is not required. It depends on the system definitions, particularly on the selection of system elements. The selection is usually made in a way that the biggest system elements possible are selected (since the complexity of the total system can be described with the lowest effort).

Furthermore, each system element should show as low as possible error modes (ideally in a degraded safe state). It is a challenge to define system components in a way that they show a low amount of possible error modes. However, the acceptable system degradations have to be clearly specified. This fact will be used in a higher ASIL as the basis for architecture development. Without barriers, which limit the number of possible error modes, such a system is no longer analyzable and the variance and the possible error propagations will no longer be controllable.

4.1 Function Analysis

A function analysis should start with function decomposition where we can see how a function is broken down from a higher abstraction level into a lower one. Figure 4.4 shows that three functions are illustrated on one system element. Level 2 shows that the functions are composed of different sub functions and also of more than just one entrance or exit. Function 1 could be a normal brake function with two activations (foot- and handbrakes) as well as two actuators (front wheel and rear-wheel brake), function 2 could be a brake activation through a sensor (for example the radar of ACC) and function 3 would be a parking brake, which is accessed by the same operational unit (foot- and handbrakes). According to that the sub functions in the blue circle would be jointly used by different functions.

Therefore, the foot- and handbrakes (see Figs. 4.5 and 4.6) as well as the front wheel and rear-wheel brake are only deployed once. If now also the foot pedal affects the parking brake, we can see that the function for both actuator have to generate different signals, or the function logic behind it has to interpret those signals differently.

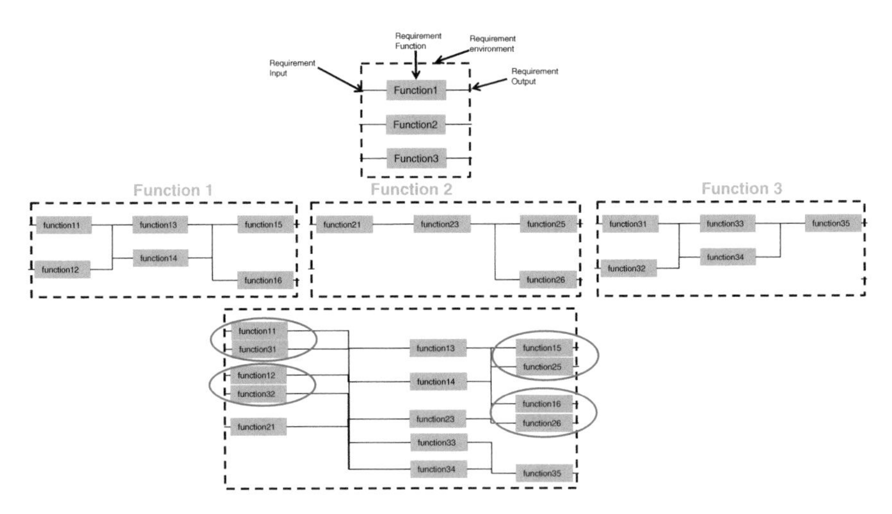

Fig. 4.4 Function decomposition

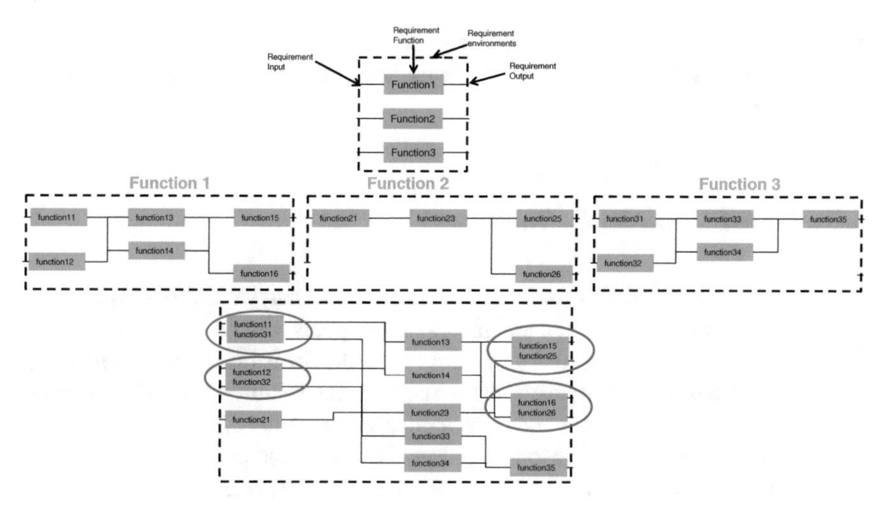

Fig. 4.5 Architectural foundation

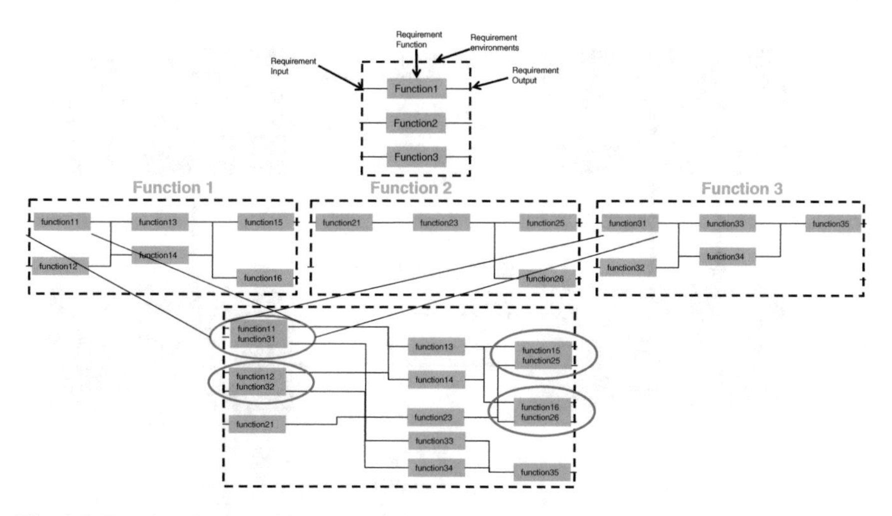

Fig. 4.6 Function decomposition and merging of functions on common functional elements

The common element now has to fulfill all requirements of all functions that are allocated to this element. This can be the maximum requirements as well as conflicting requirements. In case of the latter, two exists have to be implemented that can be activated according to the respective requirement. Those further exists differentiate themselves in the necessary details, so that the respective requirements can be implemented correctly.

Such decomposition and the consolidation in the system integration (allocation of multiple functions to one element and the development of new common functions) will have to be necessary in all system levels since only limited resources can be provided in the systems. Here we speak of a top down analysis, which later serves as a basis for the analysis of function dependencies. In order to provide a sufficient independency of target functions and a designated safety mechanism, such an analysis is inevitable.

4.2 Hazard and Risk Analysis

Officially in ISO 26262 this method is called hazard analysis and risk assessment, because ASIL, which have to be assigned to safety goals, should be a measure of the necessary activities that need to be taken in order to achieve the required risk reduction. Current risk is only marginally assessed and the aim is to examine possible situation and potential malfunctions of a vehicle system or the ITEM that can lead to a hazard. It is obviously possible to imply from the hazard and possible malfunction to the risk, but a risk assessment itself is not necessary for the application of the method.

Risk conditions in ISO DIS 26262 (see Fig. 4.7) were described similarly. This model is not complete, since parked cars can also burn down, for example, because

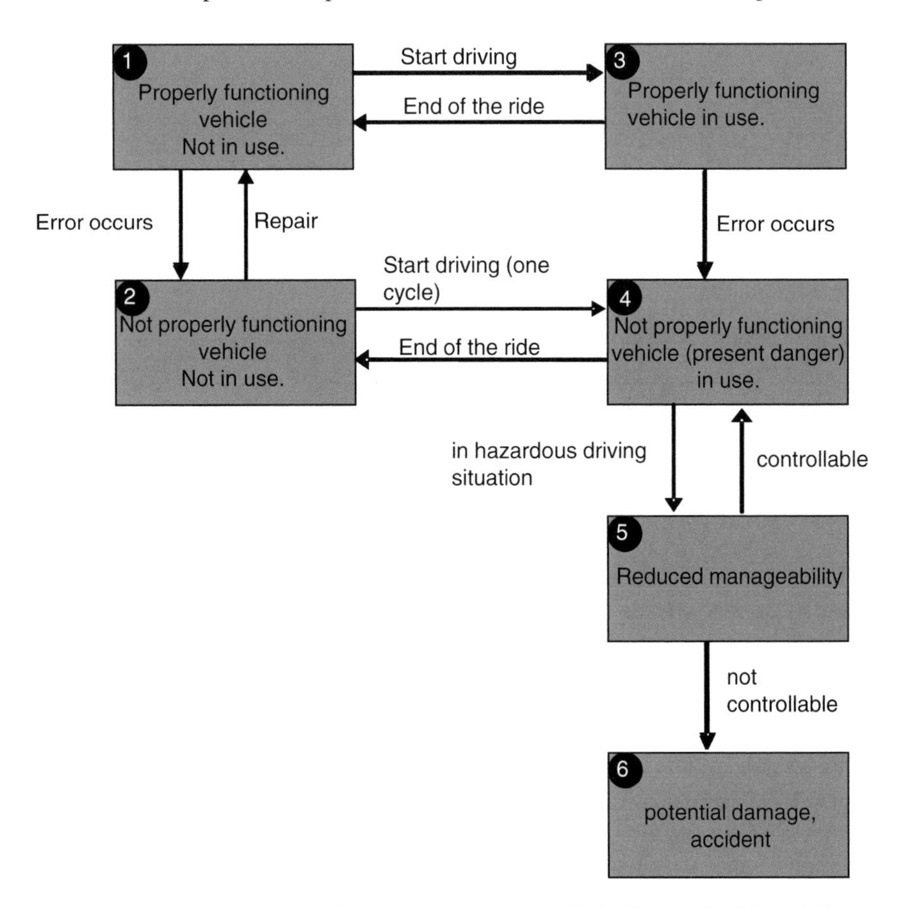

Fig. 4.7 Risk conditions as basis for the hazard and risk analysis (*Source* ISO DIS 26262)

of a malfunction of electronic components. Nevertheless, it is obvious that the number of hazards occurring from malfunctions during parking is limited. Furthermore, it is assumed that also, the dangerous situation must be present hand-in-hand with the malfunction in order to cause a hazard. If a function fails in a dangerous situation and cannot be controlled by the driver, (or somebody else endangered) an accident is likely to occur.

In other industries comparable analyses are often called "Preliminary Hazard (Risk) Analysis, (PRA)". In this context, after all analyses and verifications, the result is again illustrated through the risks of the hazard and risk analysis during the following architecture and design measures. If the illustration or the transparency of safety goals is not given, the risk analysis has to be adapted. Therefore, the analysis is never completed. The process iterations are important safety measures.

4.2.1 Hazard Analysis and Risk Assessment according to ISO 26262

ISO 26262, Part 3, Clause 7:

> ***7 Hazard analysis and risk assessment***
> *7.1. Objectives*
> *7.1.1 The objective of the hazard analysis and risk assessment is to identify and to categorise the hazards that malfunctions in the item can trigger and to formulate the safety goals related to the prevention or mitigation of the hazardous events, in order to avoid unreasonable risk.*
>
> *7.2. General*
> *7.2.1 Hazard analysis, risk assessment and Automotive Safety Integrity Level (ASIL) determination are used to determine the safety goals for the item such that an unreasonable risk is avoided. For this, the item is evaluated with regard to its functional safety. Safety goals and their assigned ASIL are determined by a systematic evaluation of hazardous events. The ASIL is determined by considering the estimate of the impact factors, that is, severity, probability of exposure and controllability. It is based on the item's functional behaviour; therefore, the detailed design of the item does not necessarily need to be known.*

ISO 26262 provides alternative approaches to enter the hazard and risk analysis:

- start with the product idea and intended functions and see the ITEM or vehicle system as a complete new development—or
- start with an impact analysis based on a previously developed product.

A systematic distinction for the vehicle system (item), which also has to be examined in the context of a system boundary analysis, is necessary for both approaches. Generally, this is required beforehand as part of the definition of the considered vehicle system. However, if the vehicle system is already partially existent, the technical characteristics and their behavior need to be balanced with the new vehicle system characteristics and their behavior and the new functional structure should be defined. In real life there aren't really any new vehicle systems in the automotive industry. Even functions such as ACC, brake assistant or park assistant and new automated functions are enhancements or just electrifications or remote-control systems of existing vehicle systems.

If now a new function (see Fig. 4.8 the green ellipse "internal logical element") is based on an existing system, the analysis of the existing system could be a challenge, due to missing specifications and unknown considered architecture it becomes quite complex.

A function that can influence a brake system and steering under certain circumstances has to examine the impact of the function on these two systems in each driving situation and in each operating situation because eventually the steering or the brake could act as an actuator for the new function. Based on that fact new potential malfunctions can occur, which should be examined by hazard and risk analysis? The previously implemented safety mechanisms of the system must not be restricted, override or invalidated by the new functions. This is why clear and complete information of the previously implemented safety mechanisms and their operating principle is necessary.

Safety goals are defined on vehicle level for the considered system or ITEM. However, various vehicle systems could influence the direction or movement of the vehicle.

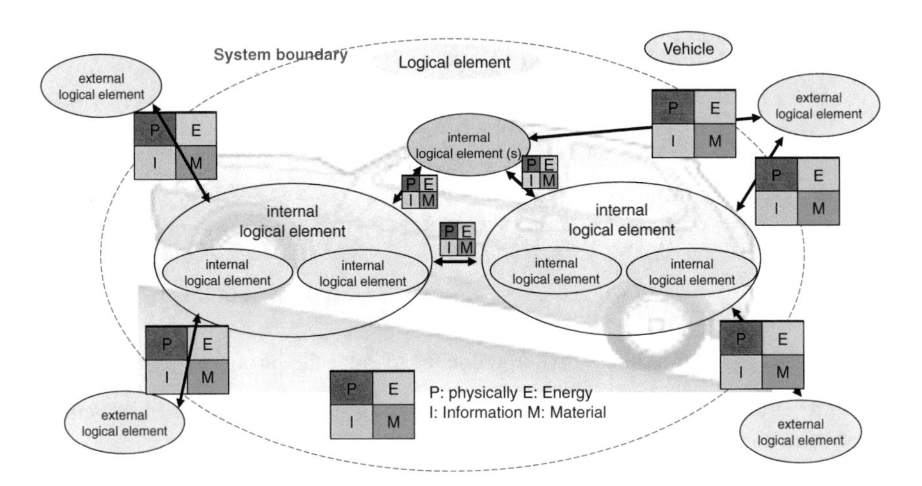

Fig. 4.8 System boundary analysis similar to *Ford FMEA-handbook*

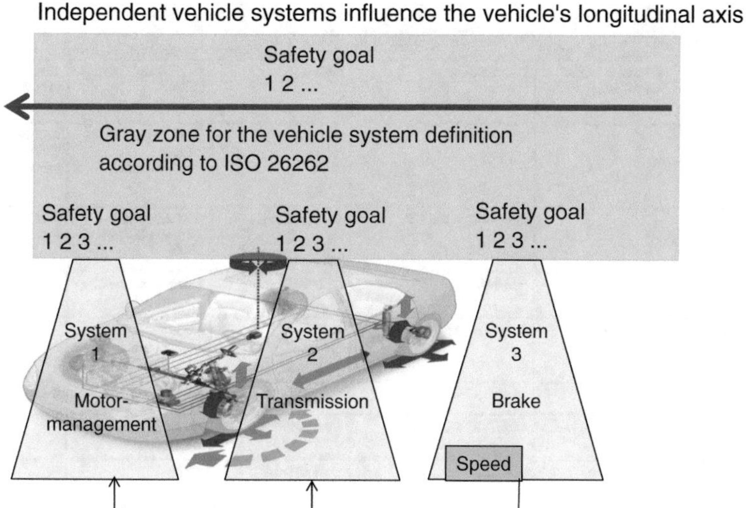

Fig. 4.9 Function levels for the definition of safety goals and the grey zone

Often common information is used in order to implement the different functionalities (e.g. speed or longitudinal movement in Fig. 4.9). This should make transparent that safety goals (and the designated ITEM) could be defined on different horizontal levels.

First, the relevant characteristics of the functions and the resulting potential malfunctions of the vehicle system should be identified. Therefore, the respective functions of the new function and the previous existing functions need to be structured and segmented.

This already creates a functional hierarchy in which malfunctions in the lower hierarchy influence functions in higher hierarchical levels. But also within the same horizontal level could influence each other. For example, a wrong information about the transmission ratio (2nd instead or 3rd gear in a manual gear system) could lead to failure in the motor management which uses the information to calculate the torque at the wheels. Another example would be a wrong sensor information lead to a wrong result for the data processing.

This means in general that there is a connected field of functions and malfunctions, which have to be considered in the hazard and risk analysis. In a further step the malfunctions together with the operating states and driving situations are expressed in a matrix.

The spreadsheet (Fig. 4.10) only shows an excerpt of the combinations, which can occur from functions and their malfunctions in specific driving situations and operating conditions. However, this figure shows an example of how complex and extensive such analyses can become if functions with various characteristics of functions have to be considered that can lead to hazards in certain combinations.

Function	Malfunction	Operational state	Driving situation	Risk scenario	Hazard / Hazardous Situation
Speed up	higher acceleration than expected	from the state	Cornering	Driving into oncoming traffic	Accident with oncoming traffic
		while Gearshift	Wet road	Getriebeschaden leads to Achsblockierer	Accident with infrastructure and oncoming traffic
		while ACC regulation	Construction trip	Drivers traveling steering wheel for effect	Accident with infrastructure
		with trailer	Column drive	Trailer destabilization	Collision
	lower acceleration than expected	from the state	Cornering	No threat	-
		while Gearshift	Wet road	Engine stalls	Collision
		while ACC regulation	Construction trip	No threat	-
		with trailer	Column drive	No threat	-

Fig. 4.10 Driving situation and operating state matrix

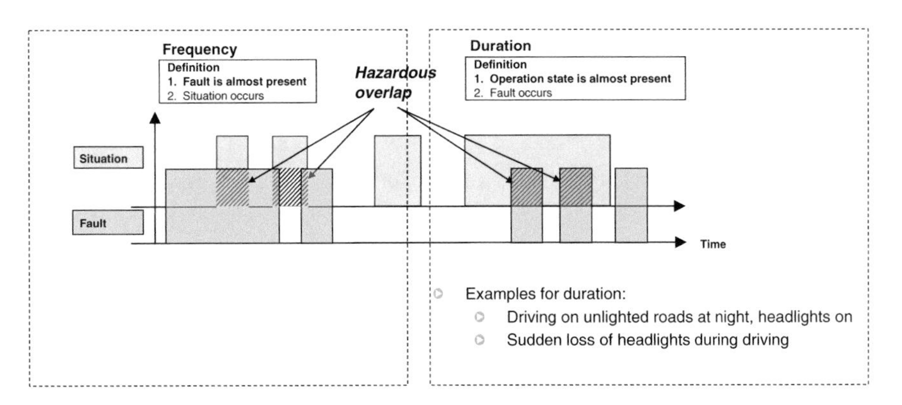

Fig. 4.11 Frequency and Permanent relation between failure functions, operating conditions and driving situations

Driving situation, operating conditions and potential malfunctions do not have only functional relations, also timing means an issue. ISO 26262 mentions two basic relations related to time (see example Fig. 4.11):

- Frequency mode: A malfunction occurs and then the vehicle gets into a dangerous driving situation
- Duration mode: The malfunction occurs when the vehicle is in a relevant operating condition or dangerous driving situation

Another possible combination is that an operating state calls a function which already has a malfunction, due to that malfunction in the a dangerous driving situation occurs. Also, these kinds of combinations and other combination have to be considered if relevant but they are not mentioned in ISO 26262.

This means that the probability of occurrence for a dangerous event is formed by the overlap of the dangerous driving situation, operating state and potential malfunction. Only the situational context is considered. Furthermore, it is important how intense the malfunction influences the vehicle in different situations and conditions. This primary affects the severity of dangerous impacts. If the stability corridor (e.g. over-under-steering or yawing of the vehicle) of the vehicle is out of its limits or an energy pulse (e.g. kinetic energy due a functional shock (sudden blocking effect in power steering leads to lateral impacts) affected an destabilization, implemented safety mechanisms are not able to control those effects after their occurrence. As a consequence such effects have to be avoided by design, so that they could not occur at all or mitigated by means of preventive detection or control algorithm etc.

ISO 26262 scales the severity of harm according to the potential injuries of passengers or other people in the area of danger. This means that the intensity or also the characteristic of malfunctions is correlates with the severity of damages. The intensity or other characteristics of malfunctions also influences other factors for the hazard and risk analyses; the controllability of dangerous situations. The term "controllability" was chosen because in this context in an automobile this mainly refers to the driver. However ISO 26262 also considers other people, which could be able to prevent a dangerous situation, for example pedestrians who can still move out of the way of a vehicle that approaches them (Fig. 4.12).

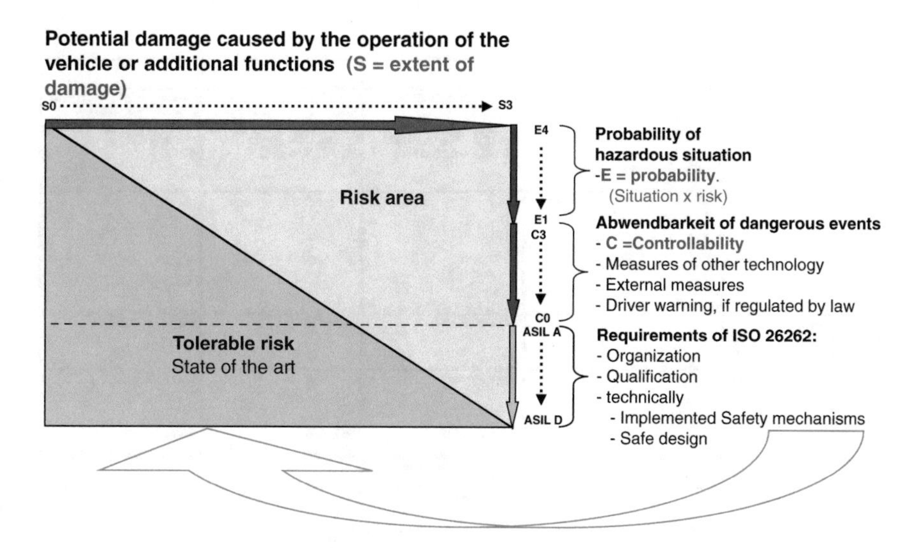

Fig. 4.12 Area of risk and tolerable risk (*Source* Various different publications)

The maximum risk is classified via the extent of damages, the severity of potential harm (S = severity)). The probability of occurrence [E = Exposure (regarding dangerous operational situations)] and controllability (C = Controllability by driver, or estimation of the probability that the person at risk is able to remove themselves, or to be removed by others from the hazardous situation) reduce the risk. The gap towards the tolerable risk needs to be covered with the respective safety measures. If safety mechanisms based on electric and/or electronic systems (E/E) are implemented for such measures, these are assigned with an ASIL. A reduction of the ASILs for EE-functions could also be achieved with measures of other technologies (e.g. a hydraulic safety mechanism).

Classes of severity (S = Severity):

A risk assessment for safety relevant functions focuses on possible injuries to people. In order to be able to compare the ultimate risks the description of the damages need to have a certain categorization. This is why we classify the severity into three different categories:

S1 > light and moderate injuries
S2 > severe/serious injuries possibly life-threatening, survival is likely
S3 > life-threatening injuries (survival uncertain) or deadly injuries

In this case it doesn't matter whether those injuries occur to the driver, any of the passengers or other traffic participants such as bicyclists, pedestrians or passengers of other vehicles.

If the analyses of the potential damages clearly reveal that only material damage and no personal damage occurred it wouldn't be classified as safety relevant function. It would be classified as severity class S0 and no further risk assessment is needed. ISO 26262 those not define further requirements for such a function, such commercial risks must be controlled by means of other measures or standards (Fig. 4.13).

Class	S0	S1	S2	S3
Description	No injuries	light and moderate injuries	Severe injuries, possibly life-threatening, survival probable	Life-threatening injuries (survival uncertain) or fatal injuries
Reference for single injuries (from AIS scale)	AIS 0 and less than 10% probability of AIS 1-6 Damage that cannot be classified safety-related	more than 10% probability of AIS 1-6 (and not S2 or S3)	more than 10% probability of AIS 3-6 (and not S3)	more than 10% probability of AIS 5-6
Informative examples	- Bumps with roadside infrastructure -Pushing over roadside post, fence, etc. - Light collision - Light grazing damage - Damage entering/exiting parking space - Leaving the road without collision or rollover			
Side impact with a narrow stationary object, e.g. crashing into a tree (impact to passenger cell)		**very low speed**	**low speed**	**medium speed**
Side collision with a passenger car (e.g. intrudes upon passenger compartment) with		**very low speed**	**low speed**	**medium speed**
Rear/front collision with another passenger car with		**very low speed**	**low speed**	**medium speed**
Other collisions		Collision with minimal vehicle overlap (10-20%)		
Front collision (e.g., rear-ending another vehicle, semi-truck, etc.)		**without** passenger compartment deformation		**with** passenger compartment deformation
Pedestrian/bicycle accident			while turning (city intersection and streets)	(e.g., 2-lane road)

Fig. 4.13 Example from *ISO 26262, part 3, appendix B*, classification of the severity factor (*Source* ISO 26262)

Classes of probability of exposure regarding operational situations (E = Exposure)

The driving or operating situation of vehicles covers from every day parking to every day driving in the city or the highway all the way to extreme situations, which ask for a constellation of different environment parameters and therefore also rarely occur. Common driving or operating situations are usually characterized by the amount of their total operating time; rare events are better expressed by their frequency.

The assessment unit E should help to categorize the various duration or frequencies. The following categories are considered for 'E':

E0 > Probability of exposure regarding operational situation is not credible
E1 > Probability of exposure regarding operational situation is very small
E2 > Probability of exposure regarding operational situation is small
E3 > Probability of exposure regarding operational situation is medium
E4 > Probability of exposure regarding operational situation is high

ISO 26262 provides in part 3, appendix B further examples for the duration and frequency

(see also Fig. 4.11 of this book):

ISO 26262, Part 3, appendix B:

Table B.2—Classes of probability of exposure regarding duration in operational situations
Table B.3—Classes of probability of exposure regarding frequency in operational situations

A typical example for the duration: A car drives by night between 1 and 10 % of its lifetime on an unlit street (Fig. 4.14).

A typical example for frequency: The average driver overtakes at least once a month.

Until today there are a lot of other publications for these categories. In order to follow the current state of the art a continuous research is required. In the course of the years a lot of evaluation and assessment will converge but also the driving behavior may change.

The probability of exposure (E) is a factor used for the ASIL ascertainment. Just like the factor controllability those two factors reduce the severity impact (S). This only applies for functions, which are part of the observed vehicle system (ITEM). The first analysis is necessary in order to identify possible malfunctions of the vehicle system related to the intended function. In order to do this the a functional concept based on new functions and the existing functions need to be structured or an hierarchical architecture need to be developed as part of the "ITEM Definition", which is inherent precondition for correct Hazard Analysis and Risk Assessment.

Class	Temporal Exposure			
	E1	E2	E3	E4
Description	Very low probability	Low probability	Medium probability	High probability
Definition	Duration (% of average operating time)			
	Not specified	<1%	1%-10%	>10%
Informative Examples				
Road layout		Mountain pass with unsecured steep slope	One-way street (city street)	Highway
		Country road intersection		Secondary Road
		Highway entrance ramp		Country Road
		Highway exit ramp		
Road surface		Snow and ice on road	Wet road	
		Slippery leaves on road		
Nearby elements	Lost cargo or obstacle in lane of travel (highway)	In car wash	In tunnel	
		Nearing end of congestion (highway)	Traffic Congestion	
Vehicle stationary state	Vehicle during jump start	Trailer attached	Vehicle on a hill (hill hold)	
	In repair garage (on roller rig)	Roof rack attached		
		Vehicle being refuelled		
		In repair garage(during diagnosis or repair)		
		On hoist		
Maneuver	Driving downhill with engine off (mountain pass)	Driving in reverse (from parking spot)	Heavy traffic (stop and go)	Accelerating
		Driving in reverse (city street)		Decelerating
		Overtaking		Executing a turn (steering)
		Parking (with sleeping person in vehicle)		Parking (parking lot)
		Parking (with trailer attached)		Lane change (city street)
				Stopping at traffic light (city street)
				Lane change (highway)
Visibility			Unlighted roads at night	

Fig. 4.14 ISO 26262, Part 3, Tables B.2 and B.3

Class	Frequency Exposure			
	E1	E2	E3	E4
Description	Very low probability	Low probability	Medium probability	High probability
Definition	Frequency of Situation: Occur less often than once a year for the great majority of drivers	Occur a few times a year for the great majority of drivers	Occur once a `month or more often for an average driver	Occur during almost every drive on average
Informative Examples				
Road layout		Mountain pass with unsecured steep slope		
Road surface		Snow and ice on road	Wet road	
Nearby elements			In tunnel	
			In car wash	
			Traffic Congestion	
Vehicle stationary state	Stopped, requiring engine restart (at railway crossing)	Trailer attached	Vehicle being refuelled	
	Vehicle being towed	Roof rack attached	Vehicle on a hill (hill hold)	
	Vehicle during jump start			
Maneuver		Evasive manoeuvre, deviating from desired path	Overtaking	Starting from standstill
				Shifting transmission gears
				Accelerating
				Braking
				Executing a turn (steering)
				Using indicators
				Manoeuvring vehicle into parking position
				Driving in reverse

Fig. 4.14 (continued)

As a result, a functional hierarchy develops in which malfunctions of the lower hierarchy influence the upper functional hierarchy. Similarly, within one horizontal functional level there are reciprocal influences. This is why a hierarchical function structuring is recommended before a hazard and risk analysis, in order to be able to describe potential malfunctions. Any changes of the functional architecture, their implemented environment and of course their characteristics, could lead to new or other malfunctions and consequently change the result of the Hazard Analysis and Risk Assessment. This is a further indication why in many industries the word "preliminary" is attributed to this analysis or mentioned in its name.

The functional concept shall consist of clearly use-cases and shall be unique and as atomic as possible, so that no functional dependencies are already part of the ITEM Definition. If so Hazard Analysis and Risk Assessment becomes very complex and the result are overlapping within the safety goals.

4.2.2 Safety Goals

According to ISO 26262 safety goals are a result of hazard and risk analysis and seen as safety requirements of the highest level. ISO 26262 indicates that a single safety goal can refer to different dangers and several safety goals could refer to a single danger. It is reasonable to describe safety goals as follows: "Avoid malfunctions (optional enhancement: in "hazardous situations", or—in "operating conditions"); the potential danger, harm or hazardous event is not mentioned. This is not always the best terminology particularly for non-functional risks, which results from insufficient robustness or robust design or inadequate design of electrical components. For example fire due to high electric current is not referable to a malfunction but to an inadequate or insufficient robust design. Fire in a vehicle is undoubting a dangerous situation. In many industry sectors voltage levels higher than 60 V are seen relevant for touch protection. In today's electro mobility risks deriving from over-voltage get more and more an ASIL assigned, so that functional safety mechanism are considered to protect over-voltage risk. In these cases for example potential causes the non-functional hazard like higher voltage, or heat are considered as malfunctions, which have to be controlled through respective safety mechanisms.

When the driving lights fail it is not necessary to name the potential danger for the formulation of the safety goal. However, it is important to differentiate if, for example, only one or both front lights fail. A malfunction, such as unintended braking of the vehicle wouldn't be considered for a light control system since such a system couldn't credibly cause such a malfunction. In the context of hazard and risk analysis the possible reaction of the driver to the failure of the light would eventually be analyzed. In this case it could be a conceivable scenario that the driver in a certain situation, out of panic, uses the brakes excessively. However, we would not try to infer a safety mechanism for the light control system from this scenario.

Malfunctions can occur in different environments, various driving situations and different performance or the impulse characteristic can have a varying influence on the driver, which can lead to more or less severe dangers. The parameters or assumptions used to define the safety goal have to be clearly specified.

Safety Goals are defined on the vehicle level (ITEM) according to ISO 26262. In this context there is no guideline to how complex or at which horizontal abstraction level is described. A safety goal can be formulated as follows: "Avoid an inadmissible pressure build-up of the brake pressure on one wheel", "Avoid an inadmissible torque build-up on one wheel" or "Avoid a defective blockade of one wheel". Basically, all three formulations could be correct. However, if we use such differing description levels for a vehicle system or multiple vehicle systems that have to be integrated into a vehicle, it could start to become confusing, since the interfaces will not match one another (compare also to Fig. 4.9).

Safety goals often describe mutual effects of possible malfunctions. A defective, far too high motor torque leads at a certain value, intensity or duration to a dangerous and for the driver, uncontrollable self-acceleration of the vehicle. A defective far too low motor torque could lead to an uncontrollable delay up until self-braking. This is why the safest function is usually a corridor, which is chosen by design limitations and/or the controllability through the driver (see Fig. 4.15).

If we consider the function "reduce vehicle speed", unintentionally or uncontrollable behavior for the driver due to malfunctions could happen. In order to fully comprehend the safety goal it is important to know how much and when which torques characteristics are considered to be correct. Furthermore, it has to be clear that after the requirement for braking, the function must have faded within a certain period of time; otherwise residual torques could again lead further endangerments. This means that the space of the correct function needs to be specified precisely; otherwise the dangerous range could not even be determined.

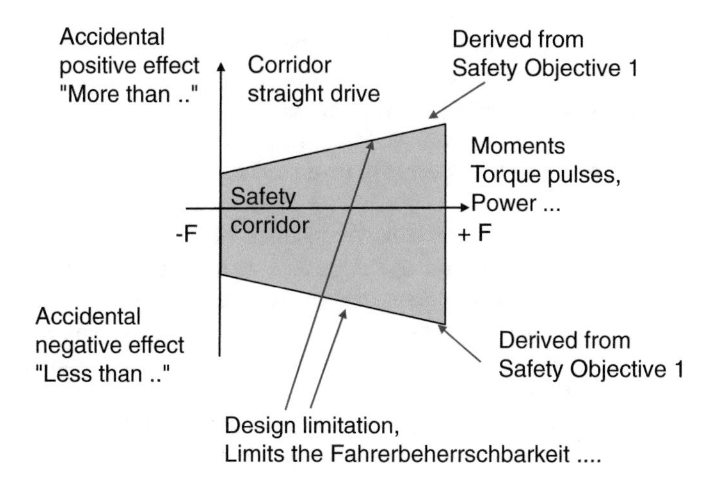

Fig. 4.15 Safety corridor of two opposing safety goals, which are derived from different characteristics of malfunctions

The alternatively a performance reduction or design limitation of the intended function change the safe corridor so that other thresholds could be tolerated. This could lead to unwanted compromises. For example; an ESC (Electronic Stability Control), which could not generate sufficient steering effect for the vehicle, could not stabilize the vehicle in case of a specific speed, vehicle weight or load etc. Besides the design limitation also the ability to control of malfunctions by adequate safety mechanism is a criterion. If a permitted driving situation or an operating condition is not distinguishable from a malfunction by the system, the intended function will need to be limited if no improved safety mechanism could be chosen. This means, that if we cannot differentiate a malfunction like drift of the yaw rate sensor from the actual yawing of the vehicle by an ESC system, no safety relevant adequate intervention could be initiated.

The brake system has to be designed in a way so that the vehicle cannot brake more or less than the respective driving situation, road surface, the current vehicle load condition etc. allows for a stabile control, the vehicle have to keep on track also in case of braking. A brake system brakes on all 4 wheels, consequently the design of brake requires a balanced braking for each of these 4 wheels. Since through the installation of the brake system itself, through aging effects, asymmetry in the weight displacement of vehicles, different road conditions, different road crib, time-delay in the data transmission in brake systems etc. as well as the current driving situation (e.g. fast carving) asymmetries of the brake force can be required or develop mistakenly. A correct specification of the permissible range of brake torque per wheel considering all possible operating modes, driving situations and possible error conditions of the vehicle could lead to extensive analysis, verifications and validations and means a development process by its own. Also legal requirements such as UN ECE (United Nations Economic Commission for Europe) R13 or FMVSS 135 (Federal Motor Vehicle Safety Standards) addressing design criterion for brake systems.

Excluding all failure combinations, which could cause such asymmetric breaking, would be even more difficult. Therefore, the easier way to prevent a violation of such safety goals could be to specify a safe longitudinal dynamics manager (controls the permitted or specified acceleration and deceleration) and a lateral dynamics manager (keeps the vehicle on track. Such control functions would prevent system errors from propagating to malfunctions which have the potential to violate safety goals.

The example shows that already the formulation of the safety goals could be essential for the deriving safety concept. A reasonable consolidation or synthesis of safety goals can reduce also the complexity of the safety concept.

The European type approval standard ECE R13 for brake systems requires for the design of brake systems, that the brake should be able to control the torque and performance of the driving engine.

With introduction of the electrification of brake systems and the consideration of electrical or electronic based safety mechanism, the requirement had been cancelled. Especially electronic reduction of motor torque in case of brake pedal action by the driver was a mayor safety mechanism to compensate the design requirement by means of implementation of safety mechanisms.

Although ISO 26262 does not address "Functional Performance", ISO 262626 requires to control the intended performance in a way that possible malfunction could be sufficiently controlled by adequate safety mechanism. If those could not be safeguarded a reduction of performance is the only safe solution.

4.3 Safety Concepts

Safety concepts are first and foremost the planning basis for the safety measures, which are the safety mechanisms to be implemented within the safety-related product and the activities to be done additionally to the normal development activities. Basically, it is a hypothesis that the implemented safety concept sufficiently safeguards safety goals. There are multiple foundations for safety concepts. Generally, the question is: Which target needs to be achieved with a safety concept. IEC 61508, 1998, in their first edition considered the achievement of a safe de-energized state in their safety goal. Formal, respective safety goals were only mentioned for the varying applications. In the beginning, in machinery engineering, only the safest de-energized state had been considered. In the oil and gas industry two typical concepts developed: TMR (Triple Modular Redundant) based on a voting (decision by majority 2 of 3) and the redundant systems based on comparison. The redundant systems were often configured to improve availability rather than maximal safety. The abrupt and uncontrolled shutting down of a refinery is indeed a more dangerous condition than a faulty state of a single process valve. In this context concepts existed from early on that use a high availability in basic electronics (often a programmable logic controller) in order to react safe and keep the system under control in any critical situation. The EGAS (E-Throttle from German VDA) concept has already been developed years ago, a basic concept for motor vehicles, which was able to control simple clear safety goals (not only the self-acceleration for which it was actually designed). Autosar primarily focused on making the application software compatible for different systems. The topic functional safety was only systematically considered after the publication of ISO 26262. Within the first versions of Autosar, the safety mechanisms are mainly reduced to the control of sequences, diagnostic handling and some principles for separation or structuring as part of the architecture.

The VDA safety concept (EGAS) which over the years has also been exported to the US and Japan, used to be the basic safety concept in the automotive industry before the publication of ISO 26262. The EGAS principle was used for varying applications. Even electric steering systems are safeguarded through a safety concept, which has been derived from EGAS. This concept here only briefly mentioned since there are a variety of different implementations for mere motor control systems. The EGAS concept is based on three levels, whereas level one carries the main function (for example the engine control), level two the functions control or monitoring level and level three the independent switch off level for level two as well as for hardware failure control.

Since this concept is based on a single microcontroller with single cores, an intelligent watchdog had been added to level three, which represents an independent shut-down path through a specific question-answer game. In the first patent it was an additional microcontroller which could even perform a limited function in case of failure of the main microcontroller. All official publication did only describe the only safety goal of prevention of self-acceleration. There was just one safety goal within a safety tolerance time interval far higher than the time constraints or performance limits of the microcontroller. The unintentional turning off of the combustion engine as a high-available safety goal had never been formulated as such. However, in engine management systems certain mechanisms are implemented, which should prevent unintended switch-off of the motor (Fig. 4.16).

Which general targets could be formulated for a safety concept. ISO 26262 clearly defined that the functional and technical safety concept should be defined derived from the ITEM Definition, a system on vehicle level and the resulting safety goals from the Hazard and Risk Analysis.

Even if the product consists of a pure software package the development should never start only with the functional and the performance customer requirements. The requirements that derived from the boundary, of the software, for example the system design, the considered software architecture, programming guidelines and the necessary integration strategy will majorly influence the software design. Also other standards like the V-Modell XT for example describes a sequential development phase before entering the V-model, themselves.

ISO 26262 also doesn't mention how a safety concept for a sensor can be developed. If that doesn't work, how could be have safety related programmable control systems (e.g. PLC, programmable logic controller) been developed? For a sensor it is often very easy: it is all about capturing the measured information and provides it correct, unaltered, directly at the product interface. In this case the detected and communicated error is seen as a safe state. The challenge is often to correctly measure the intended physical effect and converting it in an electrical equivalent. If even the measuring principle is not adequate to measure the intended physical information, even the entire system could become questionable. A pressure sensor, which tectorial membrane does no longer move upon pressure because the

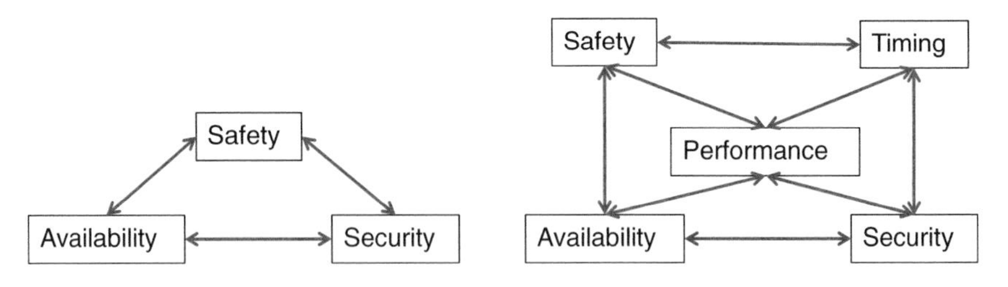

Fig. 4.16 New tension due to highly automated driving

measurement line is blocked or the membrane is indurated, is not going to be able to capture a dangerous pressure increase despite the correct electronic path. ISO 26262 does not cover performance, since its measures or requirements could not provide any solutions. Often it is possible to meet certain safety requirements through the multiple evaluations of signals, plausibilisation or also different criteria (pulse form, cosine—sinus comparison) or through homogeneous as well as divers redundancies. Even for Programmable control systems, like PLCs multiple safety goals or mechanism which reacted in different directions are challenging. ISO 26262 is applicable for high-available systems but there are hardly any examples, hinds, guidelines for the design or their realization. This is mainly because most vehicles depend on single power-supply with no redundancy, and therefore all systems need to reach a safe state in case of power-failure or a failure within the deferent systems themselves. For airplanes it is clear that they would drop from the sky without energy. Airplanes do have gliding capabilities but they are not always sufficient for a safe landing.

What about a car that runs out of energy while driving? At the start of automobile history we relied upon on the muscle power of the driver. Later, hydraulic storages or vacuum storages were used to support braking and do not overwhelm the driver. It is challenging to stop a car weighing two tons, driving 75 mph (approx. 120 kph), with sheer muscle power. It also requires a certain amount of power and concentration to steer a car without steering assistance. So far there a lot of traditional support mechanisms in vehicle engineering that don't rely on electronic energy and make car driving safer. In the future of electro mobility, we will also focus on availability, besides the aim of safe autonomous or automated driving, in order to realize such systems. What are the generic requirements and challenges of such a safety concept for future vehicle technologies? The safety goal won't be reached by simply switching off the energy. Furthermore, there will be very heterogeneous safety goals, so that being dependent on the driving situation and operating conditions we will have to switch to active of passive electronic actuators. The functions, which need to be safeguarded, will switch into different safe operating or driving conditions, once above and once below a safe corridor, depending on various influencing factors; either or, actively (with energy) or passively (without energy), a safe state needs to be achieved anyway. The problem is that we will never be able to switch to hart and fixed values, since production tolerances always require a certain tolerance margin for mass production of cars. This means that the switch off point for most dynamic safety functions needs to be learned so that the technical empirical values can be securely saved and they are clearly distinguishable from known technical errors or malfunctions. It goes without saying that the data consistency needs to be saved throughout the entire lifecycle of the vehicle. Basically, such safety concepts are based on correct and timely deterministic data processing without mutually negative influence.

In general, there is no such thing as a safe element; an element is just like evolution has formed it. Whether we look in the periodic system of elements or in the universe, we will not find an element that is by nature qualified according to ISO 26262 (or for any safety applications). Even those that we can find will be qualified or developed for a specific case of application, which by pure chance happens to be possibly applicable for the given case. Often, design decisions and also risks are not sufficiently documented, since they weren't relevant for the actual and original case of application. For example, the developer might not have known the latest state-of-the art developing principles and did not pay attention to the newest requirements and challenges, especially at a former time. It is even less possible to analyze influence factors of an unknown case of application than of one that is only partially familiar. As more we know from the development history, of the inner structure, the considered environment and the intended use of a product, as easier it is to use it also for a safety application.

4.3.1 The Functional Safety Concept

ISO 26262, Part3:

8.1 Objectives

8.1.1 The objective of the functional safety concept is to derive the functional safety requirements, from the safety goals, and to allocate them to the preliminary architectural elements of the item, or to external measures.

8.2. General

8.2.1 To comply with the safety goals, the functional safety concept contains safety measures, including the safety mechanisms, to be implemented in the item's architectural elements and specified in the functional safety requirements.

8.2.2. The functional safety concept addresses:

- *fault detection and failure mitigation;*
- *transitioning to a safe state;*
- *fault tolerance mechanisms, where a fault does not lead directly to the violation of the safety goal(s) and which maintains the item in a safe state (with or without degradation);*
- *fault detection and driver warning in order to reduce the risk exposure time to an acceptable interval (e.g. engine malfunction indicator lamp, ABS fault warning lamp); and*
- *arbitration logic to select the most appropriate control request from multiple requests generated simultaneously by different functions.*

For this section, ISO 26262 assumes that the vehicle architecture is already existent, since the safety requirements of the functional safety concept have to be implemented in the existing entire vehicle architecture. The functional safety concept used to have the basic intent to describe the necessary safety requirements, independently from their intended technical realization. It used to be obvious to allocate the requirements on functional elements. In the chapter of architecture views it is recommended to define the issuance relationship of varying elements through functional descriptions. We call such elements also logical elements. Of course also the interactions of elements among each other are functionally described. Also, the design limitations outside of the defined vehicle system have to be considered as well the geometric arrangement in the vehicle. The limitations, which result from the technical realizations of the new vehicle system, do not have to be considered in that step. That doesn't mean that it is forbidden to do so but technically it is not necessary according to the process. If it were already decided for a project, which microcontroller has to be implemented, it would be helpful to consider the microcontroller safety concept as well as the concept for basic software (including safeguarding coverage of the hardware-software-interface, HSI).

This means that it is first verified whether the intended concept is the right one for the realization before the system requirements for the electronic or software components are developed. Furthermore, there should also be a test concept, which can also show the correct implementation of the functional safety concept. According to the process we would continue but there are already findings which indicate unacceptable design limitations for the technical safety concept necessary changes need to be considered. In order to reduce safety issues that the guilty plea is documented for the product liability case for the opposing surveyor or lawyer. To reduce performance is often as inconceivable as a loss of quality or higher production costs. Therefore, the design limits are pushed to the technical limits, since ISO 26262 does not dictate any principle derating measures. Whether a microcontroller, which is run at the temperature limit or with the maximum clock frequency, actually fulfills the quality requirement of the series deviation is questionable. The situation becomes debatable, when it is known that the electric components often produce unknown technical failures at the design limits, which are not shown in any database and thus not quantifiable. For example, over and under voltage situations, which are not excludable throughout the lifetime of today's vehicles, can in this case lead to extremely critical failure combinations. The problem is that the failure behavior is only known from individual failure observations but those failures are not systematically replicable. This is why, also in the functional safety concept, design limitations should be maximally pushed to a limit of 70 % (i.e. the stack capacity should be even lower, in case it is permissible to outsource safety relevant data to the stack), so that there still are sufficient resources available in the realization process for the necessary safeguarding of failure modes, which result from the analysis (for example D-FMEA or quantitative safety analysis) for the realization details.

We use again a engine management as an example, which has to fulfill four safety goals. Some may now wonder: Why so complicated? When until now each engine management system only knew one safety goal? But nowadays it is a legitimate question whether those old wisdoms are still applicable regarding supercharged high pressure turbo engines. Since an engine management system can only formally provide so many or so little torque along the longitudinal axle of the vehicle, two safety goals would be sufficient for the vehicle level. In this case we do not assume a functional safeguarding of the thermal dangers or even fire hazard. However, there may be safety concepts that also protect the overheating of an engine or the ignition of inflammable fuel with the means of functional safety. Also, the realized safety mechanisms themselves can violate the given safety goals.

Those are the safeguarding of a mechatronic system (compare to EUC (Equipment under Control) of IEC 61508) such as a combustion engine (including for example carburetor, compressor, and turbocharger) or a fuel supply system. Considering an engine management system together with the powertrain, also a correct torque can turn out to be dangerous combined with the wrong turn rate or the motor. Therefore we can see that it is a question of the definition of the vehicle system and how the safety goals and the thereof derived safety requirements are divided and allocated to the elements. The knowledge that the engine torque and engine turn rate are not independent helps to plan the safety mechanisms and activity measures and test them in context of the verification on their effectiveness.

The processes described in this chapter should not act as a model for a functional safety concept but rather as support for considering the right aspects for the design.

Therefore we consider the following four safety goals:

- SG1: Avoid a dangerous unintended increase of the engine torque for a longer time period than t1 (ASIL C, the limit represents an array of curves, which are dependent on speed)
- SG2: Avoid a dangerous unintended decrease of the engine torque for a longer time period than t2 (ASIL C, the limit is a value dependent on the vehicle, which leads to a blockade of the drive axle)
- SG3: Avoid a dangerous unintended increase of the engine speed for a longer time period than t3 (ASIL B, the limit is a value, which leads to a self-acceleration that is not controllable by the driver)
- SG4: Avoid a dangerous unintended decrease of the engine speed for a longer time period than t4 (ASIL A, the limit is a value, which leads to a failing of the propulsion engine that is not controllable by the driver)

The limits described are variable parameter. There are for sure a lot of vehicles for which a failure of the engine management will never exceed those limits. This is why engine management systems have so far not always considered those safety goals. Also, in this context the aim isn't to define the functional safety concept for an engine management system but to consider requirements and aspects, which can practically cause challenges. The function definition or the functional concepts were derived from the definition of the vehicle system, including the correct performance results for the engine torque and engine revolution. Those are the foundation for the construction of the powertrain. Regarding this, we often see adjustments and variations for those results since for the realization there are always new design limits to consider.

Furthermore, the values always depend on the operation environment, driving situation etc. A hot engine could behave completely different than a cold engine in certain area. For some vehicles it was already discovered that an engine accelerates better at nominal operating temperature. Those are examples for safety relevant design limitations. The system can only function correctly within certain limits. Outside these limits, safety mechanisms can sometimes be completely ineffective or not operate in a timely manner.

Nowadays, high pressure is used for modern turbochargers. This pressure can sometimes increase sharply. Pressure gradients can go beyond the design limits in the short term (i.e. pulsation). An implemented pressure regulator could possibly be too slow to effectively limit the gradient. The functional concept raises the following functional requirement: Based on the accelerator pedal position and considering the current engine torque and its number of revolutions (turn rate) the throttle position and the injection pressure have to be identified in a way that the vehicle decelerate or accelerates as desired. The accelerator pedal movement represents the driver's intention.

The norm indicates that the structure should be hierarchically divided. For the four safety goals this means that the architecture of the functional concepts and safety mechanisms needs to be supplemented and transferred to a hierarchical structure.

The following figures shows how to plan a functional safety concept including the break-down of the requirements of the intended functions and from all safety mechanisms, so that the intended function doesn't need to be allocated to any safety requirements (Fig. 4.17).

The architecture of the intended function could be extended by the needed safety mechanisms and the related functional safety requirements, as required by ISO 26262. Every logical and every technical element, for which a safety requirement has been assigned, also won't be independent with its other functions. Therefore, the

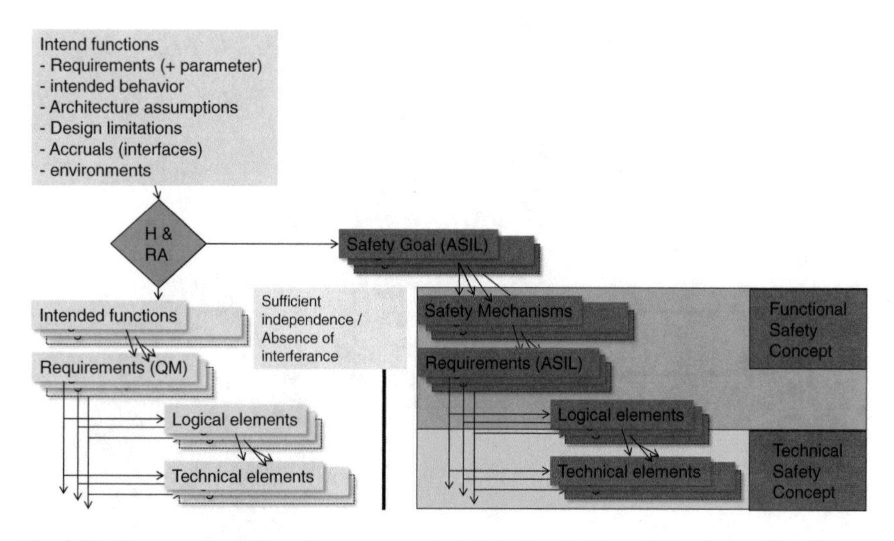

Fig. 4.17 Structuration of a safety concept, including the division of functions with different ASIL and functions without allocated safety requirements (QM)

functional and technical dependencies have to be analyzed separately for each element with regards to all safety requirements or any safety goal. This leads immediately to a high heterogenic dependency so that the effort to analyze such a system grows beyond measure and evidence of safety can no longer be provided. Therefore, the task for our engine management system is to describe the intended function and hierarchically allocate the foundation for these functions as functional architecture. Basically, the intended function will be similar for a combustion engine and an electric motor. It is a matter of accelerating the vehicle or slowing it down according to the requirements that the driver sets with the acceleration pedal. Other functionalities of a motor management such as a torque increase, in order to compensate sudden torque surges of the air conditioning compressor, or the traction control support, to better accelerate in turns or pull off on gravel are disregarded in this case. Nevertheless we need to consider the transmission, since the driver sets further requirements for the vehicle through the gear selection, as to which number of revolutions (rotational speed) and torque needs to be implemented. In the following example we assume that the transmission can digitally provide the chosen gear and we read in the acceleration pedal position as analog signal. Therefore, our system consists of the following functional groups (logical elements from Fig. 4.18):

- Logic processing
- Driver's requests detection (G, provides an analog equivalent of the accelerator pedal position)
- Motion sensor (S, provides impulses as equivalent of the vehicle speed)
- Engine speed sensor (R, provides as sinus/cosinus the engine revolution at the crank shaft)
- Transmission ration (TR, digitally provides the gear ratio)
- Safe pressure regulator (P, valve including pressure read back)
- Throttle valve engine (T, including current read back)

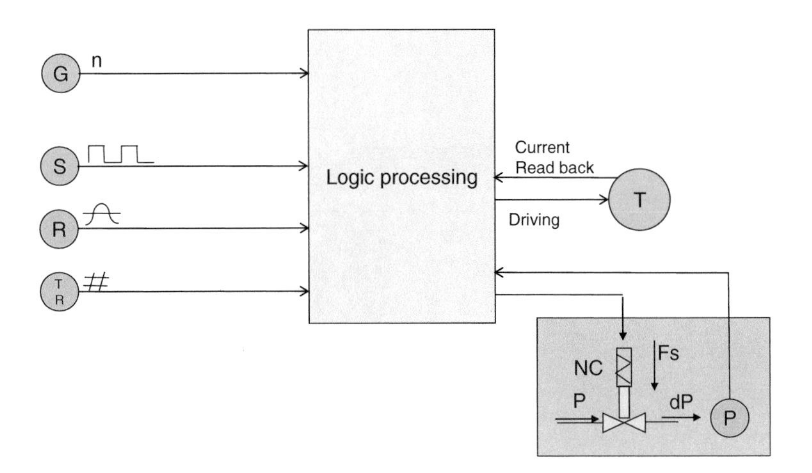

Fig. 4.18 Block diagram engine management system including the driver request

The block diagram in Fig. 4.18 already shows technical information, how logical elements can be implemented and which signals and signal types need to be exchanged between the logical elements. These assumptions should also be already documented, since in the interpretation it is already excluded or suggested that interdependencies don't exists or won't be considered at a later point in time. Thus the pressure regulator for a modern combustion engine is for sure not a simple magnetic valve, which works spring-loaded (it is probably determined through the set timing and injected with constant pressure).

We also assume that the engine fulfills a function through the throttle valve and the injection pressure, without paying attention to the camshaft or the valve injection times and that we read back a respective equivalent through the engine torque sensor.

The correct throttle valve position results from the following function:

$$\mathrm{T = f(G, S, R, TR)}$$

The corresponding parameters are shown as assumptions of the functions.

The following possible malfunctions need to be considered:

- Incorrect acceleration demand from the acceleration pedal would lead to an incorrectly higher throttle valve opening and/or to higher injection pressures.
- Incorrect delayed demand from the acceleration pedal (for example throttling back) would lead to a closing of the throttle valve or to a reduction of the injection pressure.
- Incorrect speed (Incorrect high speed could mean that MMS opens the throttle valve too far, since for high speed the trigger on the acceleration pedal would be correct for the required acceleration. Incorrect low speed could mean that the MMS injects with higher pressure, since for low speed the trigger on the acceleration pedal would be correct for the required acceleration):
- incorrectly measured values of the number of engine revolutions/engine speed would, similar to the incorrect speed, cause corresponding incorrect control of the throttle valve or incorrect injection pressures.
- incorrect transmission ratio would also lead to possible malfunctions such as an incorrect speed.

This means if we break down the intended functions into partial functions, we will always get to the same causal malfunction through the functional failure of the partial function. As a result, the data flows into the system have to be monitored in a certain tolerance band. The respective ASIL of the upper and lower tolerance bands strongly depends on the design parameters of the realization of the vehicle and the vehicle systems considered in their context. As a possible result, failures of this partial function can, with varying probabilities, lead to possible violations of the considered safety goals. If the probability of the reproduction of a failure is already excluded through the functional definition or sufficiently unlikely, no measurements are taken

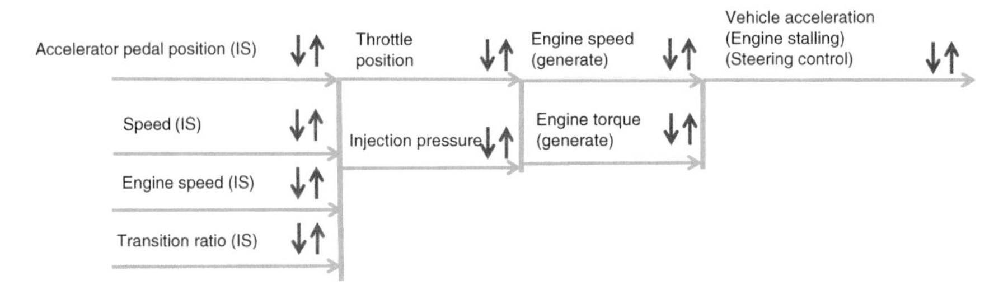

Fig. 4.19 Example for functional analysis, and possible error modes for any sub-function

in the functional safety concept at this point. Allocations on same technical elements in lower levels or at the realization can through an insufficiently robust design lead to a probability increase and those failures have to be considered in the safety analysis once again. If this horizontal process chain can be continuously analyzed and safeguarded until its function on the actuator, error propagation upwards to the safety goal could be avoided. This leads to an important rule for the analysis of error propagation: only errors that intendent affects the actuator (in some safety standards also called “final element”) have the potential to violate safety goal (Fig. 4.19).

The red arrows indicate that all parameter of functions could be incorrectly higher or lower. If the parameter drifts over time or oscillate or are sporadically wrong this could be considered in a more detailed analysis, after the first design steps and detailing of the functional behavior.

The question is, how to get to the ASIL for the corresponding partial functions and therefore over the allocation to the ASIL of the derived safety requirements?

ISO 26262 defines an iteration loop in this context so the functional safety concept will be improved as long as necessary to achieve a positive verification. Verifications are like a “repeat until loop”, unless the result is sufficient; it is typical process iteration. The verification criterion are correctness, consistency and completeness and sufficient traceability have been achieved through the vehicle system definition, the safety goals, the derived functional safety requirements and their allocation to elements of the architecture. This causes us to develop a hypothesis as functional safety concept, which should ensure that the safeguarding of all relevant safety goals. This hypothesis is based on assumptions and experiences or also on a certain systematic. It is not possible to determine the correctness in a certain horizontal abstraction level independent from the upper level. Since the safety goals represent the safety requirements on the highest vehicle level and functional safety requirements should be derived from those, breaking down the logical architecture is an effective method in order to get good results.

The functional concept may be based on the block diagram. All signals are read in with ASIL B and use the dependencies of the system function groups (logical elements) for plausibility checks to implement the decomposition or safety mechanisms. To control the throttle valve and the pressure injection we use current read back paths. It would also be possible to turn this into an ASIL decomposition, because it could be seen as a redundant implementation of a functional safety requirement. However, this would lead to further requirements, which would make the analysis far too complex. Through the current read back paths it is possible to reach a DCSPF of 99 % of all possible failures in the function blocks, for the opening of the throttle valve as well as the pressure injection. Whether the intended torques and accelerations in the realization can actually reach this diagnostic coverage is questionable for a mere functional consideration of the quality of the diagnostic coverage. This is also valid for the latent fault metrics (LFM). Since we have ASIL C as highest safety goal, the architecture metrics could be achievable through the current read back. The logic processing would also have to be reach a (Diagnostic Coverage for Single point faults) DCSPF of 97 % and a (Diagnostic Coverage for Multiple-point faults) DCMPF of 80 % regarding random hardware failures. This is absolutely achievable with a single core (microcontroller with only one processor core) if the safety mechanisms can be implemented redundantly under consideration of sufficient independence of these redundancies. Theoretically, now also all input signals and output signal circles as well as processing units can be analyzed and all error modes with a DC of 97 % could be controlled. Besides the architecture metrics, PMHF (Probabilistic Metric for random Hardware Failure) could be reached if for the basic fit rates microcontroller and components are used, which values for a conservative (as for example operating temperature below 85 °C and low junction temperature) design are derived from the usual handbook data. However, the aim is to develop safety architecture with various options for the implementation.

The reliability block diagram (Fig. 4.20) shows that we can also read back the number of engine revolutions, which as a physical quantity that could in case of an error possibly violate 2 different safety goals. With this measure we are able to directly compare the set point and the actual value for a safety system. Since the number of engine revolution depends on the injection pressure and the throttle valve, certain failures are certainly also detected for the incorrect torque status (an overly high torque leads to a steeper increase in the number of revolutions as expected). The conclusion would be, depending on the interpretation of the derived functions, that with the help of the functional architecture (logical elements) each safety goal would be safeguarded on the acceleration pedal through convincing information, with the exception of the correct reading of the driver's intention. The question is whether the driver is able to control possible arising unexpected

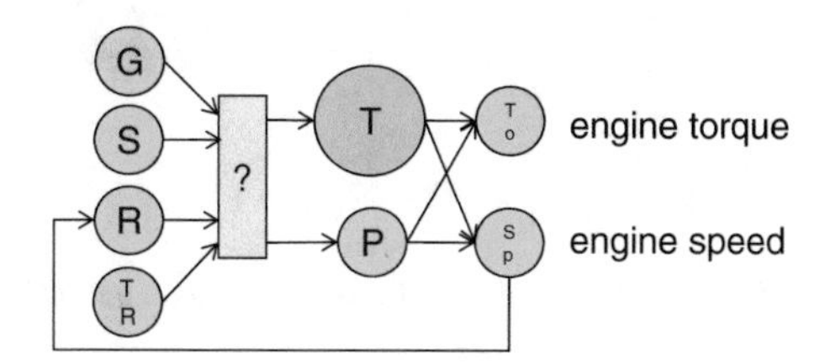

Fig. 4.20 Draft of a new reliability bock diagram for an engine management system

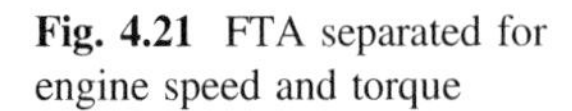

Fig. 4.21 FTA separated for engine speed and torque

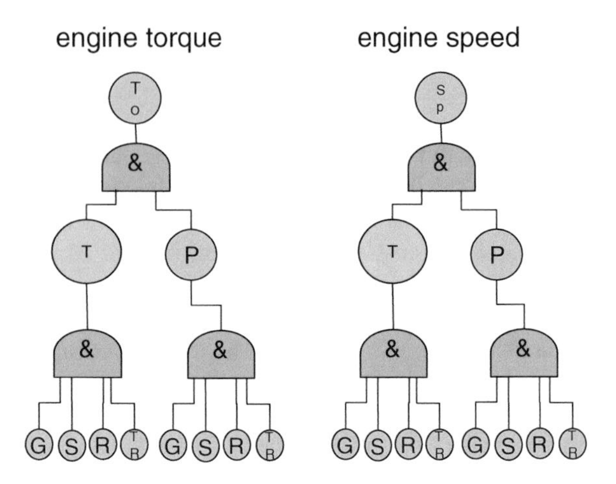

behavior or malfunctions and their performances. The answers to these questions will probably only be found in the context of the validation and therefore it should be defined as validation criteria. This is why for the constellation the acceleration pedal position will be conservatively chosen as ASIL C, according to the highest safety goals in current definition of the safety goals. For the overall safety tolerance time interval 150 ms are assumed. This is derived from the approximated reaction of the system components to the noticeable vehicle reaction. For the verification of the functional safety concept we can design a fault tree or a simplified FMEA.

In the positive view, the fault tree (see Fig. 4.21) looks very clearly arranged. If we want to know whether an incorrect, overly high speed leads to "too much pressure" or "too little pressure" we have to go through all vehicle situation and all logical combinations with all parameters in order to say which of these combinations could be possibly dangerous. This will be strongly influenced through the following realization. Because in basic development, multiple technical realizations and also multiple vehicles are often seen as an aim for the integration, the following solutions are used to mostly cover all possible options (Fig. 4.22).

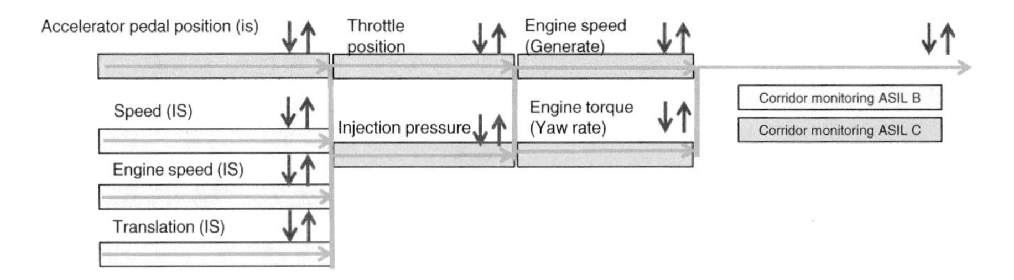

Fig. 4.22 ASIL allocation in the functional safety concept (*red arrows* indicate possible malfunctions)

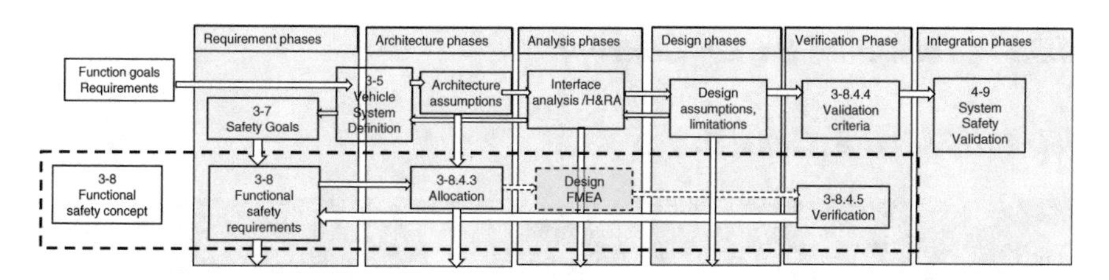

Fig. 4.23 Information flow during functional safety concept (FSC)

Since, starting with ASIL B, it is necessary to implemented safety mechanisms additionally to the functional concept, all safety requirements will be allocated to these safety mechanisms, which need to be defined additionally to existing functional requirements. It had not been considered that a safety related function could be itself designed according to its ASIL requirements, because later during implementation it will be difficult, that a function could control itself. The principle that the intended function should be controlled by an adequate safety mechanism is always safer for the realization. According to the aim of the functional safety concept now one corridor monitoring will be added for all signal flows in the functional architecture. In this context, the input signals are monitored in ASIL B and the acceleration pedal position in ASIL C. All internal calculations and status tables are logically monitored through an ASIL C corridor. Since no safety requirement has been allocated to the intended function, the "processing element the microcontroller)" has two different safety levels.

The intended function (in QM) and the monitoring function (in ASIL C) require two sufficiently independent software implementations for example two partitions in a microcontroller. The realization paths to the ASIL B sensor signals (including the sensor itself) can be realized in ASIL B. In anticipation of the technical realization the calculation for the set position of the pressure regulator and the throttle valve will also be separated. This could lead to a reduction of ASILs or at this point decomposition could already be included. However, the advantages and disadvantages should be questioned with regards to the application effort.

The verification of the functional safety concept should be supported by a draft FMEA (see Fig. 4.23) or as illustrated by a positive fault tree. A hierarchically structured FMEA would be able to support the verification very well according to the VDA approach.

In the next phase the requirements and the associated architectures and results will be passed on to the technical safety concept. In the first iterations there won't be any complete verification for sure. Therefore, the outstanding issues of the verification will also have to be passed on so that it is clear in the technical safety concept, which information are actually positively verifiable and which aren't. It is another process iteration.

4.3.2 Technical Safety Concept

ISO 26262, Part 4, Clause 6:

Specification of the technical safety requirements
6.1 Objectives
6.1.1 The first objective of this subphase is to specify the technical safety requirements. The technical safety requirements specification refines the functional safety concept, considering both the functional concept and the preliminary architectural assumptions (see ISO 26262-3).
6.1.2 The second objective is to verify through analysis that the technical safety requirements comply with the functional safety requirements

6.2 General
6.2.1 Within the overall development lifecycle, the technical safety requirements are the technical requirements necessary to implement the functional safety concept, with the intention being to detail the item-level functional safety requirements into the system-level technical safety requirements.

NOTE Regarding the avoidance of latent faults, requirements elicitation can be performed after a first iteration of the system design subphase.

The basic requirement says that the system design should be derived from the functional safety concept, whereby the architecture should still play a central role. In effect, this causes the various functions of the functional safety concept and their requirements to be again allocated to common elements. This is often the case for microcontroller.

Of course for the realization of a control system (compare to Fig. 4.24) we would need parts such as housing, a plug connector, a power supply (external periphery like the battery etc.) as well as internal components such as a printed circuit board (PCB), internal power supply and voltage distribution (internal periphery). Now we need to decide how we consider the control unit.

Do we consider a control unit including the housing and the cables or do we assort this at the first allocation? Furthermore it would be useful to consider the intended function separately from a separate software component for the intended functions and software for safety corridor monitoring. Even if we need 2 independent software elements, we have to trace the separation down into all software elements down to the software unit. This is the only way to get two independent software elements. The challenge is to identify commonly used resources and find a solution, which avoids the mutual influence of both software elements or makes their coexistence also in case of errors controllable. The example considers the following technical elements:

- Acceleration pedal sensor (P und W) consisted of two measuring devices. Once device measures the pressure on the acceleration pedal and the second device measures the acceleration pedal angle. The pressure will be transferred as a

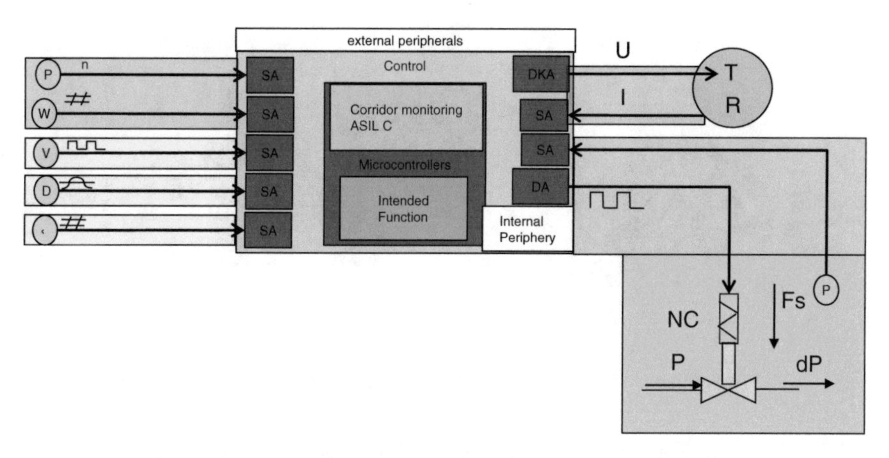

Fig. 4.24 Allocation of the ASIL attributes of the safety requirements to technical elements

16 bit digital data word. The redundant information should provide an ASIL C on the pins of the microcontroller within 10 ms.

- The speed (V) is provided as a 16 bit data word (based on impulses, converted through an external system) including the bus protection through a defined bus communication. Data should be provided at the bus interface every 10 ms.
- The number of engine revolutions (D) is transferred as a sinus and cosinus signal. This is an equivalent of the number of engine revolutions.
- The Transmission ratio (TR) is provided as a 16 bit data word (by an external system) including the bus communication protection through a defined bus communication. Current data should be provided at the bus interface every 10 ms.
- The throttle valve (T) consists of a magnetic coil and current read back. A measuring shunt of the throttle valve at the control unit should cause a measurable change in current at the control unit input pin for the current within 50 ms.
- The injection pressure (P) is initiated at the injection valve through the voltage pulse. Within 20 ms a voltage pulse at the control unit pin should fully open the valve or close it through a decline of the impulse of longer than 10 ms the valve should close. The opening can be controlled through different pulse break times. The pressure is transferred as constant current as analog signal.
- The controller block consists of internal and external peripheral elements as well as the necessary sensor and actuator adoptions and a microcontroller, which sufficiently independently provides two partitions for QM and one for ASIL C. The P2P (Pin to Pin) reaction time should be less than 50 ms.

All shown and listed parameters and variables including the redundancies and current read back function will be considered as separate technical elements and characteristics and would have to be specified with all interfaces.

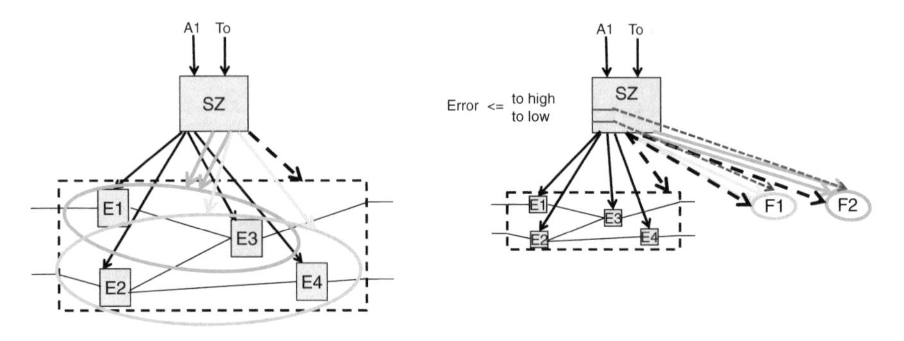

Fig. 4.25 Differentiation of functional and technical requirements

Functional and technical requirements are not different by its nature, the allocation within the architecture characterize them as such. Figure 4.25 shows that if the logical and technical perspectives are separated, the common usage of element 3 (E3) becomes transparent. We can describe functional correlations from technical as well as logical elements and we can also functionally describe the internal correlations or a technical element. This is why it is important to determine a specific description level for the technical system architecture and specify the implemented element and their interfaces. As a result the safety requirements are derived from the functional safety concept to the elements and the interfaces of the technical architecture, whereas the system interfaces do not necessarily have to be described by technical elements.

In the first process iterations of the development the technical elements are not necessarily considered. Therefore, the system elements will be described as logical elements and later more and more detailed to the technical elements, by considering the technical interfaces. It might be strange that an architecture specification actually does not start with technical information. This is why it will be a mere design decision whether an element will be part of function group F1 or F2 or considered as separate element. Therefore, architecture decision will depend on project-, product- and application-related constraints. If the components, from which the system is compiled, are developed by multiple different, cross-functional or external development teams, the interfaces should be defined according to the development teams, which are involved in the creation of the system. If product-, organizational- or project-specific interfaces are harmonized, the development will become very complex and have to be coordinated with additional effort additionally. Ultimately, this means that the technical safety requirements often refer to logical elements. In the system design the allocation to a technical elements or component only happens in further process iterations.

The interfaces become multi-dimensional; a functional, logical or technical view to an interface will provide different information about the interface. Until now, we only had to consider functional interfaces. Now, the interfaces of technical elements (see Fig. 4.26) overlap and raise new challenges or cannot implement the required characteristic by themselves but possibly together with other elements.

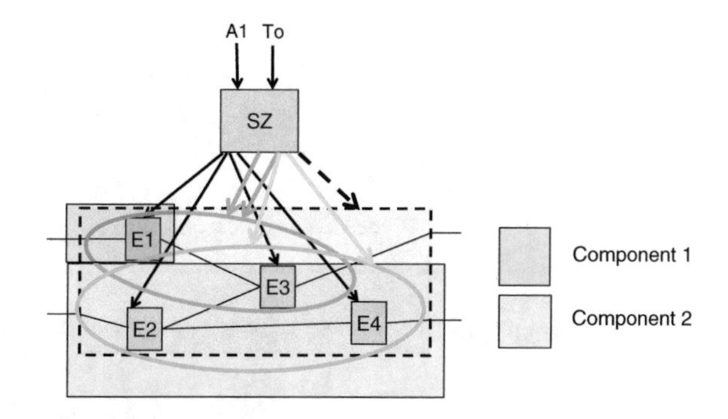

Fig. 4.26 Technical interfaces of logical and technical elements

A sensor requires, among other things, housing, a power supply and wiring. In order to read in the signal a control unit is required, which can detect the wired signal and electrically process it so that it can be read by a microcontroller. The method to derive the technical level from the functional description and specify it according to its technical requirements is not new. The literature provides this analysis for example under the name "FAST" (Functional Analysis System Technique) or also as publication of VDIs (VDI, Association of German Engineers; i.e. VDI 2803 "Functional Analysis"). Those heterogenic interfaces can no longer be understood with such simple examples. Therefore, it is important to broadly reduce and decouple those dependencies. In the first iterations of the architecture analysis we will only be able to limit it to the functional dependencies. This is why the component will only be described with logical elements in the specifications for the components supplier. The responsibility of the technical specification will be passed on to the component supplier. This is a common interface between customer and supplier. The customer provides a "functional" requirement specification, and the supplier confirms it by his performance specification (How the supplier intents to fulfil the customer requirements).

Basically, across all industries the so-called "**IPO**" principle is used for all software based systems (Fig. 4.27).

"IPO" stands for Input, Processing and Output.

Since the sensor signals and actuator control are usually implemented in extremely different ways, a signal adjustment is necessary. This signal adjustment normally happens in the **basic software (BSW)**. The interface between basic software and the **application software (ASW)** is often called a "**real time environment (RTE)**"; during runtime, this interface provides the signals, data or information channels for processing of the required software functions. In applications with a different ASILs there could be a need separate RTE for each ASIL. If the hardware software interface (HSI) is included in the basic software, it is possible to reduce the amount of interfaces within the software. However, in this case we would already receive two software components: The basic software and the application software (the processing of software functions which provide the user stake). Especially for the components with different ASILs the software components should be planned accordingly. It would be recommended that the different software components are integrated as system elements.

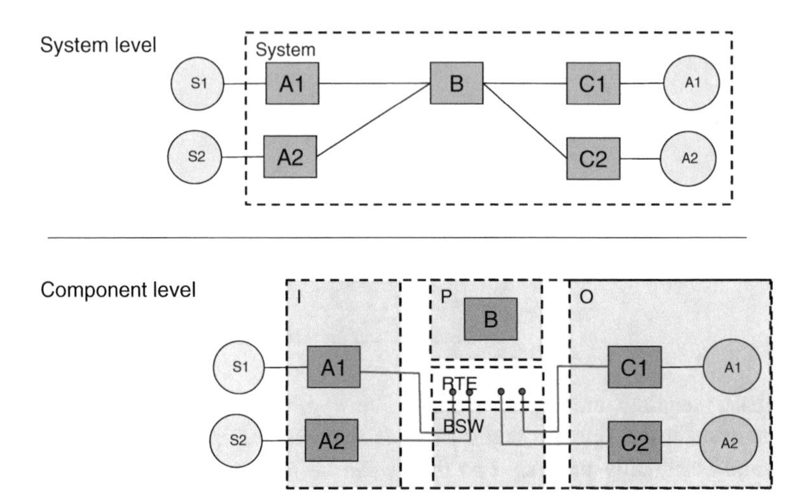

Fig. 4.27 The "**IPO**" principle derived from a system abstraction

4.3.3 Microcontroller Safety Concept

ISO 26262 does not require a safety concept even inside a component. But in order to assure a consistent development along the v-model it would be recommended. The microcontroller safety concept provides clear limitations for the implementation of different applications. Depending on the application, different safety concepts will be considered. The following aspects should therefore be already analyzed related to the characteristics of the microcontroller.

- one or multiple safety goals applicable
- event dependent safety mechanism, which are also variable based on driving situation, tolerances and systems or operating modes
- only one safety time fault tolerance interval, many time constraints, or time-critical performance requirements
- are safety- and non-safety relevant functions considered to be implemented
- are there performance requirements to be implemented as non-safety relevant functions
- is software architecture provided or are there only complexity of the partial network descriptions available (the function, which needs to be realized)
- should be legacy code or software code from foreign source be integrated
- amount of necessary partitions in order to be able to realize components with different ASIL and/or ASIL decompositions

In this context we often find a lot of indications and requirements from the definition of the vehicle system and the partial networks descriptions, which need to be realized that can exclude certain microcontroller safety concepts or at least make them appear ineffective.

The "IPO" principle could be also used for the software architecture in the microcontroller. "IPO" stands for input, processing and output. Based on this concept we now look at some basic principles for computer based safety concepts and the following fundamental questions for the microcontroller:

- How could input data provided safety conform for the application software?
- How could the functions process correct according to given safety requirements?
- How could an actuator correct controlled according to given safety requirements?
- How could the interfaces between input, processing and output be correctly safeguarded according to given safety requirements?

If we approach this according to processes, we will wonder which infrastructure is required for a safe data processing. This means that we have to create an environment, which enables us to discuss the four questions above. The environment means the architecture, design and adequate configuration of the microcontroller. Figure 4.28 of the simplified microcontroller shows the essential functional elements in a microcontroller.

The interaction as well as the functions of each element can of course be very different. However, we already have two essential groups for the safety applications. Those are functional groups, which are essential in order to put the computer into operation or initialize it. These functional groups often contribute only indirectly to the implementation of the main function. This is why they will often only be able to harm the safety function indirectly.

At the main function the signal chain from and to the pins (e.g. via port registers) as well as the **ALU (Arithmetic Logic Unit)** are involved. Different buffer or memory sections, such as the cache, are used differently in order to save data provisionally and also for the entire computing time. Generally, the program is filed in a permanent memory storage or electrically chargeable storage (flash) and then

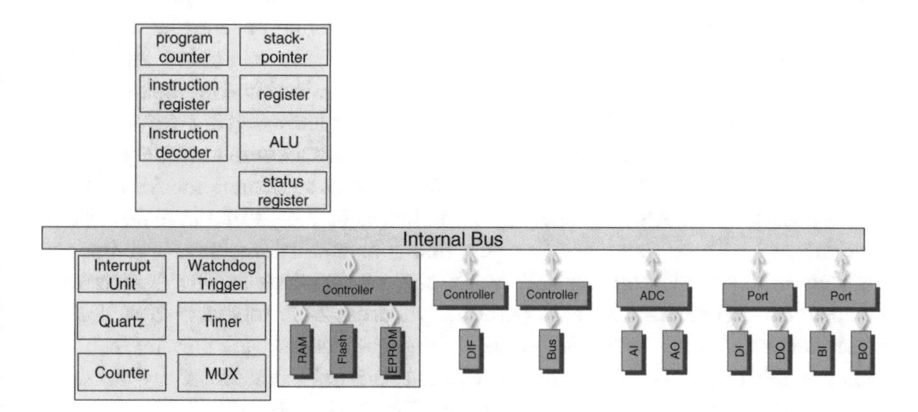

Fig. 4.28 Illustration or a simplified microcontroller

provided for the different functions in the dynamic storages or **RAM (Random Access Memory)** during computation. All register memory is used in order to reserve space for data or provide certain standardized information (status flags in flag register etc.). Counters, quartz, trigger, interrupt unit, multiplexer etc. are used depending on the programming style, compiler settings etc. in order to prepare data or control or monitor the program sequence. If those functional elements are used differently, they can also cause different failures. The more elements are used, the more error causes and data are buffered, the higher the risk that the data is falsified or processed at the wrong moment in time. Often an area approach is considered (area of total memory divided by average of used memory) for the quantification of memory errors. Even though the memory takes up large area of the silicon, the failure influence of the respective memory depends first and foremost on how often data is read in or out of the memory and in which form the correct functioning of the function is actually dependent on the data. That means errors by writing and reading of data are not depending on memory size, they are systematic errors which could not be quantified. This is why we are largely forced to control the program sequence and the data paths. The attempt to protect each individual function could lead to a positive result. However it won't necessarily be safer than a safeguarding of the entire computing function.

To safeguard each individual functional element of the computer would lead to many additional interfaces to be controlled. The number of interfaces of functional elements and considering that one functional element could, depending on the configuration, realize different functions lead to an increased number of functional variants. Since also each function shows different error modes, we will need a huge amount of safety mechanisms. Basically, this analysis will be useful, if the microcontroller manufacturer uses it and suggests appropriate safety mechanisms or safe configurations. Nowadays all big microcontroller manufacturers in the automotive market have such computers, for which the safety mechanisms are already built-in (e.g. built-in self-test) into the silicon. The manufacturer supplies respective software packages and a manual that explains how the microcontroller should be configured for the different safety applications and the respective ASIL. The question is, is this really necessary? For an ASIL D application with multiple safety goals, for which the details of the function are not yet known, it is always easier to start with a "safe" computer as a basis than to develop a computer safety concept yourself. Also some will argue to use two independent computers for ASIL D. But is this possible if an ASIL D safety function has to be discovered in a short safety time interval with big amount of vehicle functions and heterogeneous component tolerances in closed loop control? Such control functions on two asymmetric independent computers will be difficult to synchronize. In this case there are different data sets for the controller for each driving situation, each measurement and each possible position of the actuators. This data has to be synchronically processed in both computers in a certain time interval. This is why we will continue to see so called lockstep core architectures for chassis control systems more often. Lockstep computing means, for which two controllers cores processing the same software code and their result will be compared. Until ASIL C the VDA safety concept

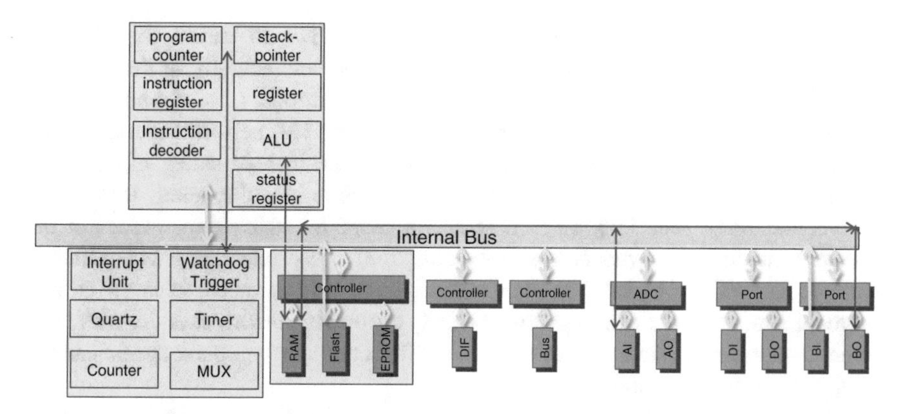

Fig. 4.29 Signal chains considered in a microcontroller

(EGAS) used to be a good solution to safeguard vehicle functions. The involvement of sensor signals and the actuator control have often not been sufficiently considered. Separation of the 3 different levels is realized in many different ways over the years the approach has been implemented.

Based on a simplified functional model for the computer, with consideration to the two different functions, which need to be safeguarded, we will now illustrate a safety concept applicable until ASIL C. Since for an ASIL C function we already have to corroborate in the software, before we control an ASIL C action, a certain redundancy of the sensors needs to be present. Generally, it is possible to say that a single analog signal cannot be safeguarded more than up to ASIL B. For the actuator control often a current read back path ensures that we can verify the information for the control through the microcontroller.

For this simple example (Fig. 4.29) the analog signal would be provided for the logic solver through the **analog digital converter (ADC)**. From the Logic Solver the control information will be provided for the output by digital (usually a more bit binary output) to a transistor and the transistor opens a valve. According to this description only a few functional elements are directly involved in such a safety function.

In early microcontroller such as an Z80, even a multiplication function had already been buffered in RAM or these multiplications already stand for multiple arithmetic operations we can see that the real functional elements for a single heterogeneous operations are not possible to be determined. How the different core (or even from ALU) operations are used by different compiler settings, and what memory areas are used, is normally unknown. The approach, which many interpret from IEC 61508, suggests simply protecting all functional elements with the DC for the corresponding integrity level. This would be safer than to do nothing but it is certainly not efficient and not sufficient. In this case probably half of the computing power is used for the protection and safety mechanisms and the intended control function will be realized at the limit of its runtime. Therefore, for simple safety systems, it won't be useful to protection all functions of the microcontroller with the highest ASIL. It is also significantly more problematic to ensure that the SW configuration uses the microcontroller functional element actually in the intended way. Usually we reach the limits of today's microcontrollers, for a complete sufficient safeguarding the resources are not sufficient. The performance is greatly restricted and it will no longer be possible to ensure that all safety mechanisms actually work efficiently within the safety time requirements (respecting time-constraints); especially since it is not predictable, which error modes of the different functional elements actually cause the safety relevant effect. Often a stuck-at is well protected but data falsifications, masquerades (data in correct protocol frames or data formats but factually incorrect), incorrect addressing, different data order or wrong buffering are only detectable by computer specialists but often critical for the application.

When the application should be ready and it is realized that certain errors can be injected and that they are not controllable or transferable into a safe condition within the safety time requirements, compromises have to be made, a bigger computer has to be chosen or often at the costs of the performance, the safety functions need to be prioritized.

Nowadays, more and more small microcontrollers are implemented in sensors, which are then used for the filtering of data, the linearization or digitalization.

Also in this context a detailed analysis of the target function and the required safety measurements is necessary in order to avoid an overall generalized safeguarding. Almost all error modes could be the possible cause of failure for safety relevant effects. To safeguard the computer just like **programmable logic control (PLC**s) would not be appropriate. This is why a safety concept isn't only useful in regards to the vehicle systems but also for function or system elements, which later have to be integrated in the vehicle system.

4.4 System Analyses

System analyses represent methods of the system theory. Depending on the abstraction comparable methods can be applied to any systems. Even in order to investigate sociological dependencies, decompositions are used in order to describe characteristics of groups of people, to analyze or to classify.

The subject of the analysis is often a model or a restricted image of reality.

This means that the system has to first be described at a certain abstraction level within the considered context and the expected behavioral patterns.

Generally, we also speak of deductive and inductive analyses, whereby the general induction infers from the details to the commonalities and the deduction aims to explain the details from the commonalities through certain premises or general conclusions. In order to take those terms in the context of a technical system analysis, we have to get back to speaking about a horizontal abstraction level in the system structure.

In the context of ISO 26262 we start with the general abstraction level (higher abstraction level), which often describes a system at the vehicle level. The deductive safety analysis now has the task to verify, based on a safety concept and the determined safety goals, a hypothesis, which identifies characteristics (positively seen) or malfunctions (or their causes) that can negatively influence the safety goals or higher level safety requirements. Furthermore, appropriate measures are required that could prevent, mitigate, avoid or reduce these influences. The inductive safety analysis is based on characteristics or their potential fault causes. In this context it is investigated whether these potential failures could violate safety goals at the described correlations or ways of propagations in the system.

4.4.1 Methods for the System Analysis

Historically, NASA is often mentioned, which described **FMEA ("Failure Mode and Effects Analysis")** methodology first for the failure analysis for the project Apollo in 1963.

In 1977 Ford introduced an alternative, the fault tree analysis as well as dynamic event trees, to the automobile industry. In Germany this alternative was described in DIN 25 448. NASA adopted it shortly after and it was soon also introduced to other industries.

However, failure analyses have been carried out long before that. The first failure analyses were based on methods brought to the US from German scientists after the Second World War but also in other industries some methods developed and became standards. This chapter will also cover reliability block diagrams (which can be certainly directly derived from Lusser's law) as well as HAZOP (hazard and operability) analysis, which has its roots in the chemical industry as well as the oil and gas industry.

FMEA (Failure Mode and Effect Analysis)

The classical Design-FMEA considers mechanical components and its aim was the adequate design and resulting characteristics. From an exploded view of a component the detail characteristics had been evaluated and the consequences of certain deviation from nominal or defined tolerance had been analyzed (Fig. 4.30).

The automobile associations like VDA and AIAG have described the essential methods in this context. Standards had been improved in other industries based on their requirements. FMEA is according to ISO 26262 an inductive method for the safety analysis. However, all FMEA methods in the automobile industry are widely based on the sequence of failure cause, failure and failure effect. The kind of measures to improve the product or avoid, mitigate errors, or their propagation had been defined and applied differently in the standards. The evaluation factors of failures are called as follows:

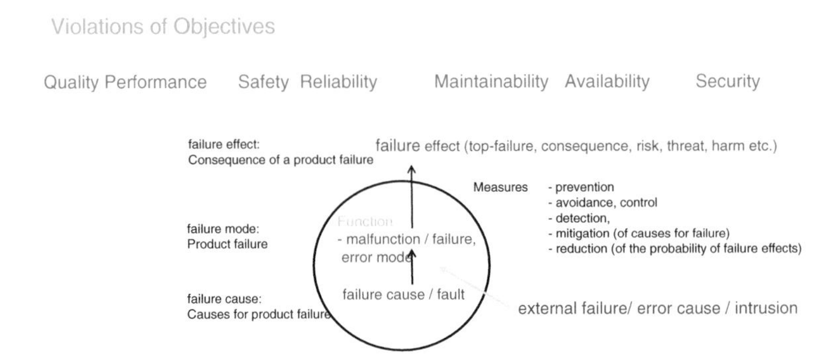

Fig. 4.30 Basic principle of (inductive) failure analysis

- Severity of damages (S)
- Probability of the error occurring (O)
- Probability of the error detection (D)

The severity-class of a FMEA is generally determined by the failure effect. This severity (S) is defined differently than the severity of a hazard and risk analysis. Generally, it can be said that the severity of the hazard and risk analysis refers to human impacts or harm (and environment damages in other standards). The severity-class in a FMEA refers more to the vehicle itself. This is why the vehicle is often used as the root element of a VDA-FMEA. In a hazard and risk analysis also the driving and operating conditions are considered, which would lead to very heterogenic structures for a FMEA. This doesn't mean that it couldn't be applied for simple systems. The probability of the class for faults occurrence (O) and the class of detection of faults (E) are generally based on the assessment of the failure cause. Those three factors build the **risk priority number (RPN)**. The two factors S and O (SxO) are often combined to the so-called criticality. However, it is often useful in the different FMEA methods, to assess the factors separately. The probability of error propagation is often not considered in the classical FMEAs.

VDA already developed a hierarchical concept 20 years ago, which requires 5 steps for the analysis.

The typical failure analysis itself only happens in the third step. Steps 1 and 2 are analyses and information, which are needed in FMEA in order to present the object of analysis or are other analysis by themselves. Steps 1–3 could be seen as the illustration of a deductive analysis, since functions and structures are broken down or decomposed (Fig. 4.31).

Fault Tree Analysis (FTA)

The fault tree analysis is a key method for the development and analysis safety relevant systems in almost all industries and also for nuclear power plants or the aerospace sector. With the help of the fault tree analysis it can be examined, in which combination, by using Boolean logic, relevant elements could fail and cause

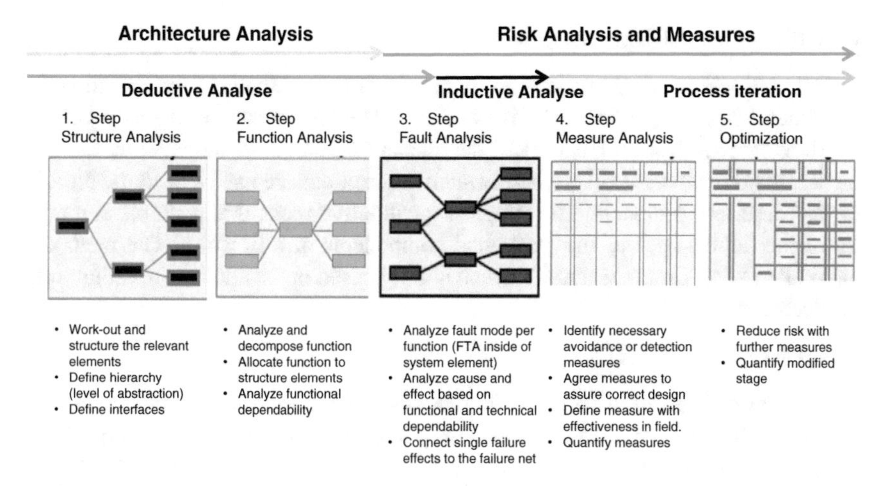

Fig. 4.31 FMEA in 5 steps (*Source* similar to VDA 4)

undesired states or events, like a failure of an engine. The aim of the fault tree analysis is to determine the minimal amount of events, which can cause such a top event and therefore to detect specific weaknesses and also unintended states in the system.

The historical background of the fault tree analysis comes from the military sector. At the beginning of the 1960s this technology was first used by the U.S. Air Force and then spread to other areas of the aerospace as well as the nuclear energy sector.

Efforts are made to illustrate and analysis more and more comprehensive and complex systems as fault trees. Fault trees are based on Boolean logic, which can be investigated on minimal cutsets with various algorithms with different targets. Special forms of the cutset-analysis are Quine-McCluskey for the minimization of the Boolean logic, the MOCUS algorithm, and the algorithm of Rauzy on binary decision diagrams, the algorithm of Madre and Coudert with Meta products, and the search strategy CAMP DEUSTO. Those algorithms are based on the different data structures and procedures for the determination of minimal cuts.

The Boolean algebra and graphic fault trees generally build the foundation for such analyses. The fault tree analysis is seen in ISO 26262 as deductive analysis, whereas this cannot be said for the higher analyses of the cut-sets. These analyses are also not necessary for the requirement or architecture development at the descending branch of the V-model; they would more support the analyses of ISO 26262, part 5, Chap. 9. This is the ascending branch of the electronic V model considers already a first realization of the product. The here required analysis and the related metric like (PMHF, Probabilistic metric of random hardware failure) targets to identify error propagations based on failure rates for random hardware faults and their potential to violate given safety goals.

Reliability Block Diagrams (RBD)

Reliability block diagrams are, similar as the fault tree analysis, are considered in ISO 26262 as example for deductive analysis. The blocks can be logically put into relations through Boolean algebra. If the blocks are quantified, the relations can also be described mathematically, whereas such descriptions are used as a foundation for formal description models. The simplest quantitative method is a simple summing up of the failure rates of the individual components of a function. The method is also called “Part Count Method”, which simply based on an addition of failure rates of electric parts.

Event tree analysis (ETA)

ETA also evolved differently in the automobile industry in the context of company standards. In most cases the aim is to complement driving situations in the system FMEA, which however, can very quickly lead to a very complex illustration. In order to assess certain top failure in different driving situations, ETA can be a useful illustration. In this case there is often an overlap with the hazard and risk analysis. In the old DIN 25419 the illustration of ETA was described first and foremost but not in kind of way, which for example it is possible to infer a certain dangerous event from a specific failure behavior. The newer DIN EN 62502 (VDE 0050-3): 2011 responds more to the methodology. However, it describes a different methodology than typical automotive standards considers.. In this new standard also the combination to FTA and the reliability block diagram is also described. The representation method is also different than the symbol description in DIN 25419. An essential point, which DIN EN 62502 (VDE 0050-3): 2011 addresses is the determination of an analysis area. With this illustration (see Fig. 4.32) we can see that the relation of failure, operating situation and driving situation for electronic products could easily become very complex.

Markov analysis

The Markov analysis is primarily used to assess the transition from one condition to another. The formula for the safety architectures in the informative part 6 of IEC 61508 was derived from such models. Those formulas are gladly used for the EE safety architectures. However, the basic principles and assumptions, under which the models were designed and the formulas derived, are often not known or not applicable for the realized automotive architectures. Often only one failure at the time is assumed, therefore ageing affects, error combinations, dependent, transient or latent failures are not derivable from this formula. For approximations or as help for the quantification these formulas are applicable if corresponding further analyses are applied.

HAZOP (Hazard and operability studies or analysis)

HAZOP is a qualitative analysis of the hazard potential of failure conditions or malfunctions of individual technical elements. In interdisciplinary teams, i.e. architect, system analyst or tester, the target function (also the design appropriate

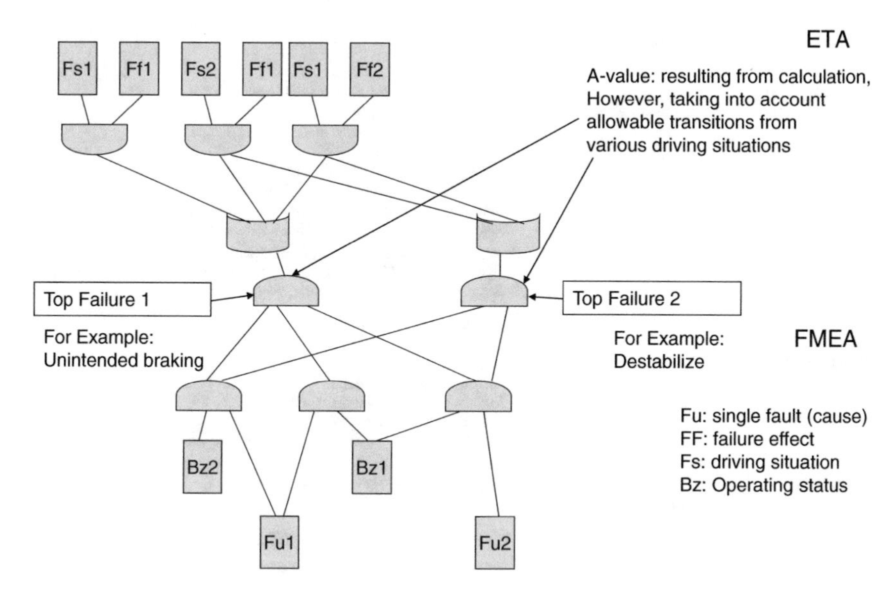

Fig. 4.32 Combination of Event Tree Analysis (ETA) and Failure Mode and Effect Analysis (FMEA)

functionality) is inferred from the detailed description of the analyzed object as well as the possible failure functions or failure behavior and measurements from a structured questioning of the target function.

Similar questions (see Fig. 4.33), just like those for HAZOP, are also considered in the fault tree analysis in order to find causes of known failure.

SAE had founded a working group and published a standard concerning detailing of the HA&RA according ISO 26262. Considerations for ISO 26262, ASIL Hazard Classification, SAE J2980 [5] Prop Draft F2011ff. In this standard the key or guide words from other industry HAZOP had been derived to automotive applications (Fig. 4.34).

This standards provides further very important background information, which help to perform Hazard and Risk Analysis for automotive applications.

4.4.2 Safety Analysis According to ISO 26262

Safety analysis is the term with which ISO 26262 describes methods such as FMEA and FTA. Neither in this book nor in ISO 26262, was the intention never to define these methods anew. ISO 26262 mentions in part 9, Chap. 8 the different methods and lists general requirements in the context of ISO 26262 for these methods. In the individual development of items, system or components the safety analyses are invoked in the respective context, based on the specific requirements for this method.

Guidewords	Importance	Aspect
NO, NOT, NOT	negation the goal function	no part of the goal function is performed, but it also happens nothing
MORE	quantitative growth	Physical size: weight, speed temporal aspects: too late, too early Covers: too late, too short, too high, too low
LESS	quantitative decrease	Feature: Material, dynamic, thermal conductivity Dynamics: heating, pressure build, move, rotate
AS WELL ... AS	qualitative growth	Function object is achieved, side effects occur as - Dynamic effects: heat resistance increase capacity reduction, overshoot - Tangible effects: contamination, wear, Corrosion, fire Contact
PARTIALLY	qualitative decline	Partial feature - Performance is not reached - Vibrations (signal always breaks) - Information or incomplete signal - Sub-functions or sub-elements without function
REVERSAL	Negation of goal function	Direction, sign, action principles
DIFFERENT TO	Operating states	Ignition cycle, sequences, state machines, Memory organization in the microcontroller, Generation of data fields

Fig. 4.33 HAZOP—key words, meanings and aspects

→ GUIDEWORDS

System Function Vs. Guidewords	No Activation	Incorrect Activation 1 (More than requested)	Incorrect Activation 2 (Less than requested)	Incorrect Activation 3 (Wrong direction)	Autonomous activation (Request is zero)	Locked Function (Failure to release)
Electric Steering Assist Function	Loss of Steering Assist	Excessive Steering Assist	Reduced Steering Assist	Steering in the opposite direction	Unintended Steering	Locked Steering
Brake by Wire (base brake functionality)	Loss of Brakes	Excessive Brake Apply	Reduced Brake apply	-	Unintended Brakes	Stuck Brakes
Stability Control Function (ESC with brakes)	Loss of ESC	Excessive Yaw Moment Correction	Inadequate Yaw Moment Correction	Incorrect Yaw moment Correction	Unintended ESC Apply	Stuck ESC

Source: SAE, J2980 Revised Proposed Draft F 2011

Fig. 4.34 HAZOP—guide words adapted to automotive applications

In part 9 there is only one indication for the differentiation of the deductive and inductive safety analysis. The inductive safety analysis is described as “bottom-up” approach. It is considered that known causes of failure and their unknown failure effect are examined. ISO 26262 mentions Failure-Mode-and-Effect-Analysis

(FMEA), Event-Tree-Analysis (ETA) and the Markov (modelling) analysis as examples for inductive methods.

The deductive safety analysis is accordingly describes as “top-down” approach, which examines the unknown causes of failure from known failure effects. ISO 26262 lists Fault-Tree-Analysis (FTA) and Reliability-Block-Diagrams (RBD) as examples for deductive methods.

Generally, it is only possible to conduct a functional analysis as a top down approach since the function of a certain hardware element needs to be known first before the cause of failure can be derived. On the other hand, for the failure observation of sheer technical (realized) elements, such as components or structural elements, the failure of elements and also random hardware failures can be inferred from the characteristics of the elements. The failure effects can then be referred to as inductive analysis. However, as a result only the inductive analysis truly addresses the effects of random hardware failure. By a deductive approach only requirements for random hardware-failure could be elaborated so that required failure rates and related diagnostics could be specified. By adequate verifications and tests the fulfillment of given metric requirements could be shown. This will widely indicate a mix of inductive and deductive analysis, whereas the lower level is designed inductively and in the upper level the technical malfunctional behaviour is described functionally, so that the consistency (see Figs. 4.35 and 4.36 analysis phase) could be assured.

A second topic of inductive and deductive safety analysis, we also differentiate between qualitative and quantitative safety analysis. The quantitative safety analysis should also consider the frequency of failures, but for both the fault modes and effecting errors need to be analyzed. Generally, the norm says of course that the quantitative safety analysis is used to fulfill the quantitative metrics from part 5, Chaps. 8 and 9.

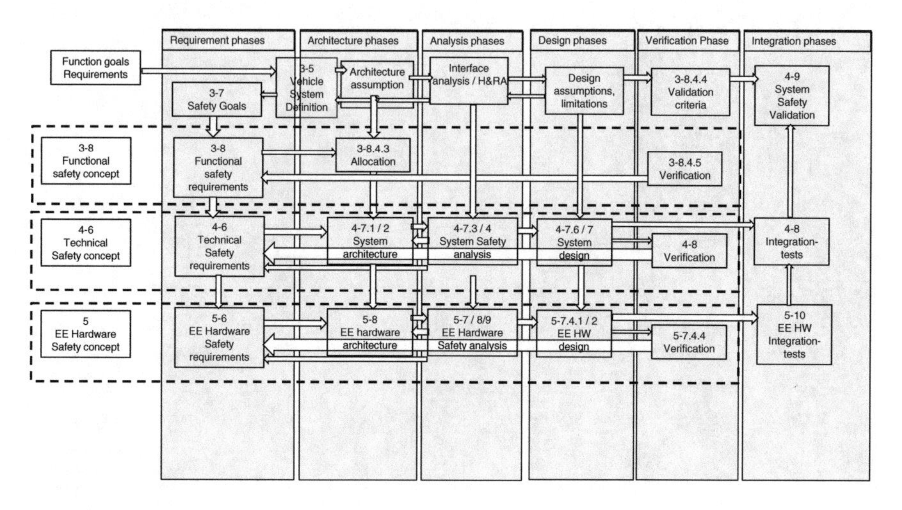

Fig. 4.35 Information flow in system and electronic hardware development

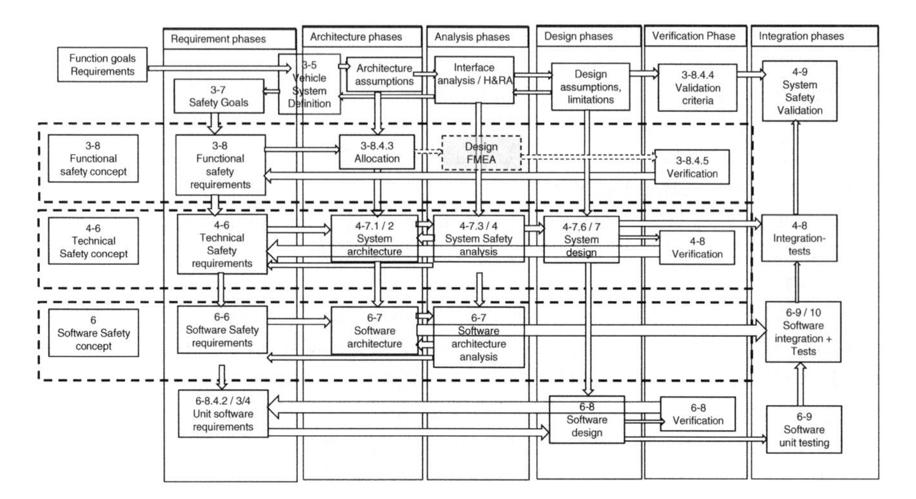

Fig. 4.36 Information flow in system and software development

The following qualitative safety analysis methods are listed:
ISO 26262, Part 9, clause 8:

8.2.2 Safety analyses are performed at the appropriate level of abstraction during the concept and product development phases. Quantitative analysis methods predict the frequency of failures while qualitative analysis methods identify failures but do not predict the frequency of failures. Both types of analysis methods depend upon knowledge of the relevant fault types and fault models.

8.2.3 Qualitative analysis methods include:

- *qualitative FMEA at system, design or process level;*
- *qualitative FTA;*
- *HAZOP;*
- *qualitative ETA.*

NOTE 1 The qualitative analysis methods listed above can be applied to software where no more appropriate software-specific analysis methods exist.

8.2.4 Quantitative safety analyses complement qualitative safety analyses. They are used to verify a hardware design against defined targets for the evaluation of the hardware architectural metrics and the evaluation of safety goal violations due to random hardware failures (see ISO 26262-5). Quantitative safety analyses require additional knowledge of the quantitative failure rates of the hardware elements.

8.2.5 Quantitative analysis methods include:

- *quantitative FMEA;*
- *quantitative FTA;*
- *quantitative ETA;*
- *Markov models;*
- *reliability block diagrams.*

NOTE 2 The quantitative analysis methods only address random hardware failures. These analysis methods are not applied to systematic failures in ISO 26262.

The note 2 refers already to the problem with random hardware faults and systematic failures. The cause of an error in a system is never only related to random hardware faults, the cause for a random hardware fault is often already a systematic fault such as wrong selection of the part, wrong estimation of environmental impacts, production errors etc. That means all quantitative methods rely on systematic analysis, where the quantification could only consider as an indication or as a metric for comparison or balancing of the architecture or design. In other standards those approaches are considered also as semi-quantitative analysis. Furthermore, the question is if these methods are only the kind of representation of the result of the analysis rather than the indication for the analysis itself.

The way how the methods are applied based more on the context of the process how parts 4 (system) and 5 (electronic hardware) of ISO 26262 considers the application of the methods. The descriptions in the previous figures (Figs. 4.35 and 4.36) show the information flow for a system and the electronic hardware as well as for a system and the software. In this context we can see how ISO 26262 invokes the safety analysis and where the results will be applied. This illustration is not showing a complete information flow; a complete illustration can only be shown in regards to a specific product development and its realizations. Depending on the maturity level, for example in a B- or C-sample cycle of the development, very different iterations of the information flow have to be considered mainly because of modifications and changes of the product.

Basically, the inductive and deductive safety analyses are invoked in the architecture related chapters of ISO 26262, in which the inductive analysis is often demanded for all ASIL requirements and the deductive analysis only for ASIL C and D safety requirements.

In the system development, deductive and inductive safety analyses are invoked in order to investigate the demand for safety measures to avoid systematic errors or faults. Furthermore, there are requirements to use quantitative metrics from part 5 as criteria for safety measures, which are effective during the vehicle operation, which mainly means implemented safety mechanism to control systematic errors or faults. These safety mechanisms and their efficiency could only measures by using the reference to metrics for random hardware failure. Therefore, for ASIL D each a deductive and inductive safety analysis needs to be performed and one of these safety analyses has to be at least quantified in order to assess the system architecture (architectural metrics) and to investigate the probability of the violation of the given safety goals (PMHF, Probabilistic Metric for random Hardware Failure) or the second method based on limitations of failure classes (see ISO 26262, Part 5 clause 9.4.3 (Evaluation of each cause of safety goal violation).

In the electronic hardware development the inductive and deductive analyses invoked in part 5, Chap. 7.4.3, safety analyses. The norm requires in this context especially the qualitative analysis of the cause-effect relationship. Further, the error causes and effectiveness of the safety mechanisms have to be proven for the avoidance of single- and multiple-point faults. In addition to that the correct design of the electric components or their sufficient robustness is required in the following Chap. 7.4.4 as part of the verification of hardware design. In this context there is also a back reference mentioned to the previous Chap. 7.4.3, since this is generally supported through a Design-FMEA in the automobile industry. This means that traditionally we choose a classical risk based approach, whereas ISO 26262 additionally requires a complete verification with regards to all relevant electronic requirements.

In other words, if there isn't a sufficient independency between parts or function groups within hardware components, which aren't a part of the realization for the considered function group or considered element of the safety relevant functions, have to be considered for the design verification as well. It seems to be a similar analysis as later required as "Analyses of Dependent Failure", but the requirement is relevant for all ASIL.

Software architecture analysis

ISO 26262, Part 6, Clause 7.1:

7.1 Objectives

7.1.1 The first objective of this sub-phase is to develop a software architectural design that realizes the software safety requirements

7.1.2 The second objective of this sub-phase is to verify the software architectural design

7.2 General

7.2.1 The software architectural design represents all software components and their interactions in a hierarchical structure. Static aspects, such as interfaces and data paths between all software components, as well as dynamic aspects, such as process sequences and timing behaviour are described.

NOTE The software architectural design is not necessarily limited to one microcontroller or ECU, and is related to the technical safety concept and system design. The software architecture for each microcontroller is also addressed by this chapter.

7.2.2 In order to develop a software architectural design both software safety requirements as well as all non-safety-related requirements are implemented. Hence in this sub-phase safety-related and non-safety-related requirements are handled within one development process.
7.2.3 The software architectural design provides the means to implement the software safety requirements and to manage the complexity of the software development.

The Tables 4 and 5 of the chapter provide following recommendations for safety mechanism, which are classified for different ASIL.
ISO 26262, Part 6, clause 7:

7.4.14 To specify the necessary software safety mechanisms at the software architectural level, based on the results of the safety analysis in accordance with 7.4.13, mechanisms for error detection as listed in Table 4 shall be applied.

Table 4 provides the following depending on the ASIL classification:
Table 4: Mechanisms for error detection at the software architectural level:

- Range checks of input and output data (all ASIL)
- Plausibility check (++ for ASIL D)
- Detection of data errors (for all ASIL)
- External monitoring facility (++ASIL D, +B and C)
- Control flow monitoring (++ASIL C and D, + for B)
- Diverse software design (++ASIL D, + for C)

ISO 26262, Part 6, clause 7:

7.4.15 This subclause applies to ASIL (A), (B), C and D, in accordance with 4.3: to specify the necessary software safety mechanisms at the software architectural level, based on the results of the safety analysis in accordance with 7.4.13, mechanisms for error handling as listed in Table 5 shall be applied.

NOTE 1 When not directly required by technical safety requirements allocated to software, the use of software safety mechanisms is reviewed at the system level to analyse the potential impact on the system behaviour.

NOTE 2 The analysis at software architectural level of possible hazards due to hardware is described in ISO 26262 5.

Table 5 provides the following depending on the ASIL classification:
Table 5: Mechanisms for error handling at the software architectural level:

- Static recovery mechanism (for all ASIL)
- Graceful degradation (++ASIL C and D, +A and B)
- Independent parallel redundancy (++ASIL D, a for C)
- Correcting codes for data (for all ASIL)

The rules for the usage of such tables are given in Chap. 4 of all parts from ISO 26262. The rule says that any table has an introducing requirement; only this requirement is a requirement to be fulfilled to claim compliance to the standard. In this case the measures or hints in the table give recommendations, but the need of implementation in this case derives from the software architectural analysis. The aim of the software architecture analysis should be to show, which measures are necessary according to ISO 26262. The verification of the architecture should then indicate their effectiveness and correctness or ability to control the error modes. In this context it is important that the focus lies on the safety analysis of the software architecture and not on an analysis, which is applied for SW-Units. In practice this means that the analysis doesn't consider the internal structures, calls and realizations within for example a C-file. Furthermore, it is also assumed that this software safety mechanism is implemented in the architecture level and therefore it should also control the dedicated error modes on this level of the architecture.

If we consider these recommendations we often end up with separate C files, which realize the desired functionalities and others that realize the safety mechanisms. By using partitioning (separation of functions and functions monitoring by safety mechanisms), which can also already be implemented in the system architecture level, the complexity can be strongly reduced. A mix of safety relevant functions and non-safety relevant functions at the software design level, which is inside of a SW unit, is actually not considered in ISO 26262. The SW-Unit would have to be developed according to the highest ASIL. This could then lead to an issue with ASIL D, since at this point redundancies or an implementation of diversity software is recommended.

4.4.2.1 Failure/Error Propagation

ISO 26262 considers an error propagation, which is defined through the three terms "Fault, Error, Failure".

In a FMEA (compare Fig. 4.37) fault can be assigned to the cause of failure, an error to a failure type and failure to failure effect. If we distinguish the causative level, failure level and failure effect level also as different horizontal abstraction levels, we easily fulfill the requirements of ISO 26262, part 9, Chap. 8. In this part it is required that the safety analyses are oriented at the architecture, just like the FMEAs.

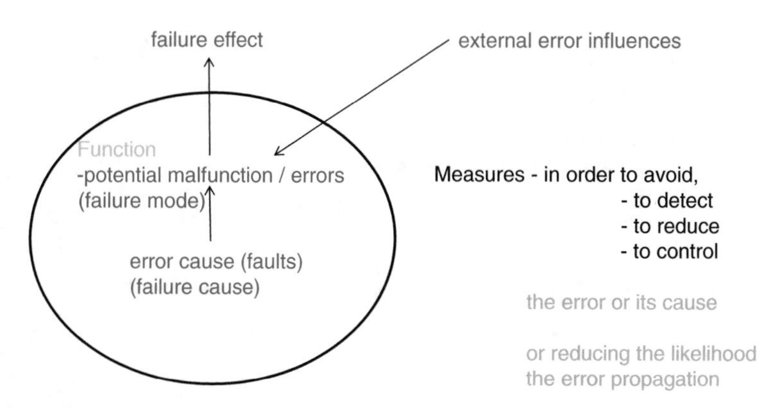

Fig. 4.37 Error propagation in a 3 step FMEA approach

The figure shows the correlations with the FMEA, whereas the term of the function is already considered differently. The failure effect in the FMEA is often seen outside of the scope, for example for a system the impact for the vehicle.

In doing so, the system itself realizes with its components functions, which then in combination or through the interaction with other systems or components in the vehicle performs the vehicle functions. A brake system is always dependent on the wheels of a vehicle; without a correct functioning of the wheels a brake cannot function correctly. This is why the functions according VDA FMEA are allocated to each error class (also type of failure). This means there is one function for each error class. Generally, the failure level is seen as the level in which a product (system, component, and element or object to be analyzed) is specified. Therefore, it is also possible in this context to test against the requirements, which have to be founded through the implemented functions.

The diagram in Fig. 4.38 shows that the probability of the error propagation strongly depends on the design of the signals, distances, dimensioning and environmental influence factors. Whether corrosion leads to failure dependability on the various influencing factors and the fact that high currents could also clean contacts, consequently it could prevent corrosion at contacts, but man should not rely on it always. Corrosion can also lead to extensive contact resistance and thus to a temperature increase or even to a fire in the control unit. Whether low current, which operates the windshield wipers, can actually control a throttle valve or maybe prevent its closing is questionable but generally such effects cannot be excluded, if not EMC requirements is fulfilled for all elements within a vehicle. The example (see Fig. 4.38) also shows that depending on the position of the observer, the respective levels of the failure classes can be described differently. As a reference point we can always use the specification, since in context of its verification the aim is to show that the own specification is correct, meaning all observed, measurable or calculable requirements are implemented correctly at the product boundary level. Through negative tests (for example injection of faults) correct behavior can be tested in a failure situation or a stress test can test the design limitations or its

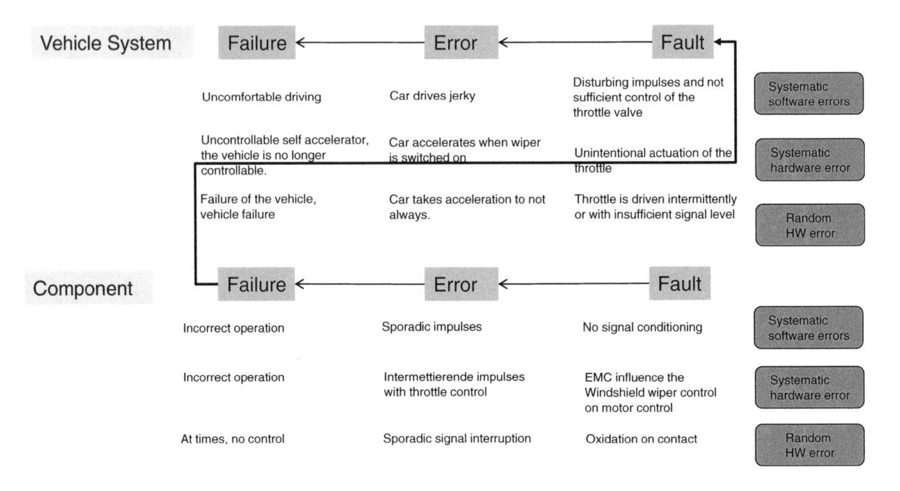

Fig. 4.38 Examples of an error propagation through multiple levels (*Source* Inspired by ISO 26262, part 10)

robustness. This means if mathematical proof can be provided that the signal to noise ratio for the windshield wiper control meets the EMC requirements and therefore no unauthorized interference occurs in the neighbor signals or electric devices. The examples in the error propagation in Fig. 4.38 won't be seen as real cases since the calculations are based on wring assumptions.

The cause of failure has a strong influence on the failure characteristic and it is often not easy to recognize the reason. However, if it is possible to limit the cause, further possible measures can be found, which make it possible to control the failure. These analyses refer more to the classical Design-FMEA, which questions whether the characteristics of the analyzed object meets the non-functional requirements e.g., quality. This method is used to answer questions in the mechanical sector such as: "Is a M6 screw suitable to safeguard the construction?" or in the electronics sector whether a 100 ohm resistor is the right choice.

FMEA is primarily used to find the right and necessary (risk based approach) tests for the design verification.

The classical Design-FMEA focuses more on problem oriented correlations (compare Fig. 4.39) than architecture analysis and different horizontal layer of abstractions.

Often, it is based on the Japanese method "**5 Why**", which says that the cause of the failure needs to be detected after asking "why" at least five times. If we assume a defect, then this defect doesn't necessarily need to have a negative impact on safety. However, such a defect could of course, depending on the perspective of the user, lead to a limitation in the applicability. Small noises can cause a car buyer to withdraw from the purchase. Furthermore, permanent failure can lead to a different error pattern than sporadically occurring errors or drifts, which can cause a critical failure behavior through a certain transient. The nature of the cause of failure can go

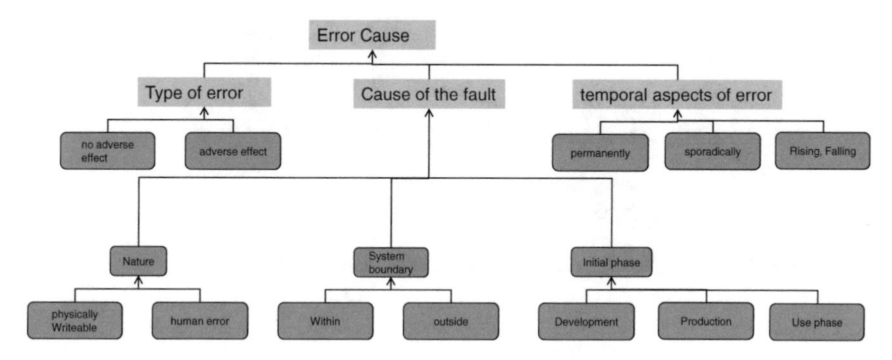

Fig. 4.39 Alternative correlations of a cause-and-effect relationship

back to an unknown or incorrectly judged physical influence but an influence can be also incorrectly assessed by an (human) error. It is even more difficult if the designer (person deciding on detailed product properties) or the reasons for the design decisions are unknown. It is true that the detailed design decision is often transparent at a higher abstraction level but most of the time, which influences through the decision are considered to be uncritical and which are relevant are not communicated. The same applies for design decisions of other vehicle systems or design influencing activities in the production of components. Particularly for e-mobility we can see that the user behavior of the vehicles is often simply unpredictable. Which function or which components are active over which period of time, with which intensity and how the ageing effects affect the components is often hard to foresee. Due to the low amount of comparable systems in the field, any field-effect could have different causes. This shows that the failure correlations are difficult to assess, because we do not know what kind of stress happened during the use of the component. If we look at the requirements in ISO 26262, part 5, Chap. 9, which say that the failure correlations that lead to a violation of a safety goal have to be quantitatively assessed, we can imagine that this will be extremely challenging. PMHF (Probabilistic Metric of random Hardware Failure) in part 5, Chap. 9 consider only the random hardware failure, but the decision, on how these random hardware failures can actually effect the safety goal is widely determined through the above described influence factors. This is why the results from the quantifications for the architecture metrics should be developed very conservative. Failure combinations are quantitatively widely determined through the probability of the common cause or common mode effects. If it refers to an independent failure, the probability will most likely go towards infinite and if we look at dependent failure, the degree of dependence will determine it (basis principle of Kolmogorov's axiom).

However, the degree of dependency should, according to ISO 26262, not be quantified since there are no generally known methods or principles, with which these dependencies can be quantified.

Basically, errors can be avoided, the probability of occurrence can be reduced or also the error propagation (see Fig. 4.40) can be avoided or their probability

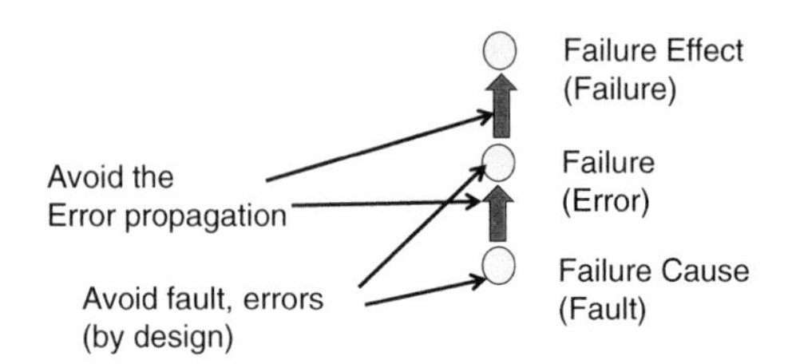

Fig. 4.40 Error propagation in relation to failure effect and causes

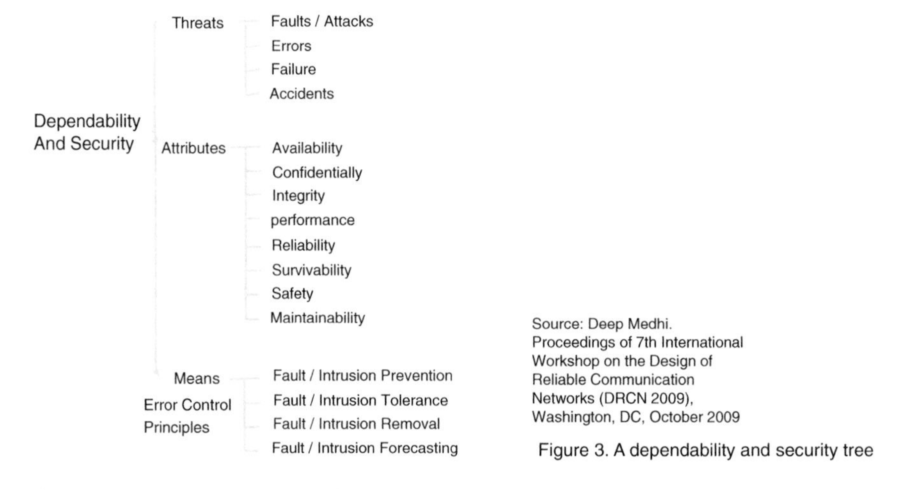

Fig. 4.41 Dependability and security models (*Source* Deep Medhi [3])

reduced. Avoiding or reducing the failure probability happens through the sufficiently robust design and the avoidance or reduction of error propagation through the architecture. For the error propagation we distinguish between the error propagation on one level into the higher abstraction levels, for example from the components through the system to the safety or security goal or the error propagation within a horizontal level (Figs. 4.41 and 4.42). The error propagation principles is not only applicable to typical dependability issues like safety and reliability it could also applied for security by using the approach as shown in Fig. 4.42.

The error propagation within a horizontal level can affect the following relations:

- from one element to another element (for example between transmitter and receiver),
- from one input to another output (if an input of a transistor is wrong also relating outputs could be wrong, even in case of correct processing),
- through incorrect data entry of configuration data or operation modes incorrect output values can occur despite correct processing.

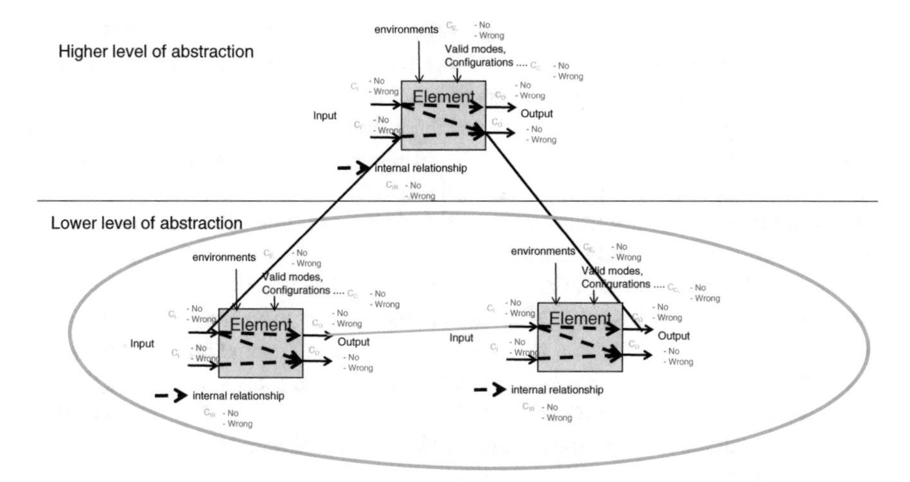

Fig. 4.42 2 level of horizontal abstraction including potential error causes

- through inadmissible environmental conditions incorrect output values can arise (microcontroller produce wrong random output reactions at the output in case of overheating or increased EMC).

Furthermore it is important for the error propagation in a higher abstraction level how errors can, for example occur in different operating situations. In this context the fault tree analyses considers different type of cut-sets.

Birnbaum or Fussel-Vesely importances describe how errors and situations are related and how high the overlay is for failure combinations. In this context the following aspects are to be mentioned:

- Is the situation or operating condition already existent when the failure occurs (for example during close loop control) or do we get into an operating condition or situation and the failure is already existent (for example a switch off channel does not work in a case of failure)?
- Failures occur sporadically and other failures are only dangerous at a certain moment in time when a intermediate error occurs.
- Failures are only dangerous in certain situations or operating conditions but also cannot be controlled in all situations and operating conditions.

These dependencies of failures and the possibilities of the error propagation have to be analyzed so that if needed, appropriate measurements can be taken.

4.4.2.2 Error Propagation in the Horizontal and in the Vertical

Considering the horizontal levels and different perspectives we can also speak of an error propagation in the horizontal levels (see Fig. 4.42) and of an error propagation upwards to the safety goal. Through the inheritance of requirements and the horizontal structure, natural systematic failures are also propagated downwards (for example from the system to the component). This is primarily accompanied by a functional analysis and leads in context of the failure analysis relatively quick to an infinite complexity. Overheating and the consequences in a microcontroller are not possible to forecast, it is just a probabilistic distribution. In many safety standards the horizontal propagation of errors are related to safety integrity measures, which are also considered as safety barriers.

The possible error propagation in the horizontal level can also only be limited conditionally, since each systematic failure at the specification can for example be a potential source of error. In order to achieve something like more error-safeguarding appropriate barriers are needed in the horizontal levels, which can prevent further error propagation. IEC 61508 proposed in its first edition not to mix safety-related and non-safety-related software or software of different integrity levels in one microcontroller.

Other safety standards require that even separate control units are used, since especially the environmental conditions can also influence the electronic functions and with separate control units it can be assumed that the functions in different control units are not dangerously influenced by environmental influences at the same time in the same way. It is mainly if fail-operational or fault-tolerant functions are to be considered, but also extensive heat or EMC could lead by common mode effects to unexpected behavior even if safety mechanisms are correctly implemented.

ISO 26262 allows different ASIL for software in one microcontroller, and also having legacy software, software which have not been developed according a safety-standard or software from foreign sources in a sufficient separated environment. But except, to perform an adequate "Analysis of dependent failure" the standards provide no guidance. How to design fault-tolerant or even fail-operational architectures and designs and how to deal with such horizontal barriers are not considered in ISO 26262.

Therefore, the question is how to make sure, for example, that the software in a microcontroller does not get negatively influenced by the surrounding hardware or the functions, which need to be added in later design phases? In order to reach sufficient safety all the way to the highest safety integrity levels, it is common practice nowadays to integrate redundant controller-core in lockstep mode. Even if the redundant electronic components are used also in power supplies, overall on printed circuit boards and in common wiring harnesses, this independency can be reasoned through a sufficiently robust design. However, if we are looking at a highly available safety requirement, the requirement for an independent energy supply, which ensures that in the case of malfunctions can still be implemented, is inevitable. Through the analysis of dependent failure (especially the common cause and common mode analysis) it is possible to point out on the hardware side that the chosen components are sufficiently robust in the specified environment so that the hardware has no negative influence on the redundantly implemented safety functions. This means that all single faults, which through the environment can negatively influence the correct function, are designed so fault-tolerant that this influence can be ruled out. If it is possible to rule out certain failure through a sufficiently robust design of the system to the requirements, which arise for the environment and the design limitations, only the error propagation will be relevant through the functions within the horizontal level. (Example: A sensor delivers the wrong value at the input and therefore the microcontroller calculates correctly an incorrect value at the output).

For functional recursions, such as for example close-loop-control, this analysis can already lead to a random complexity.

A close-loop-controller could be considered by standard functional elements (see Fig. 4.43). The control feedback could be influenced by any of the elements of the entire close-loop-controller. The correct function of the controller cannot be monitored by a comparison of the input and output conditions. Errors at and between the input and output, at the permitted configurations as well as the defined environment, can lead to errors in the reference variable, the control deviation, the manipulated variable, the disturbance variables, the control variable and the feedback and consequently to failure of the close-loop-controller (Fig. 4.44).

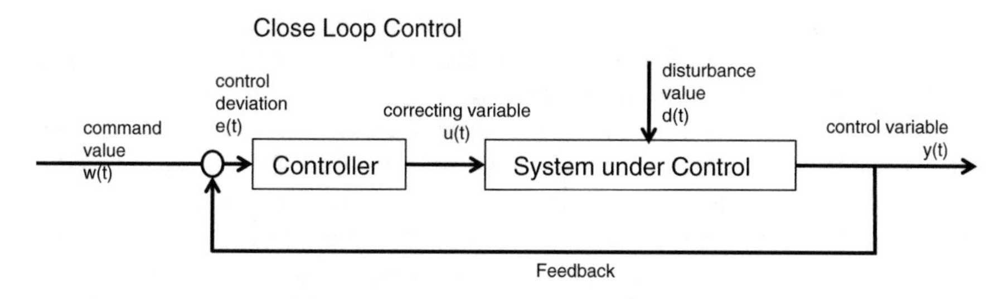

Fig. 4.43 Typical closed-loop controller

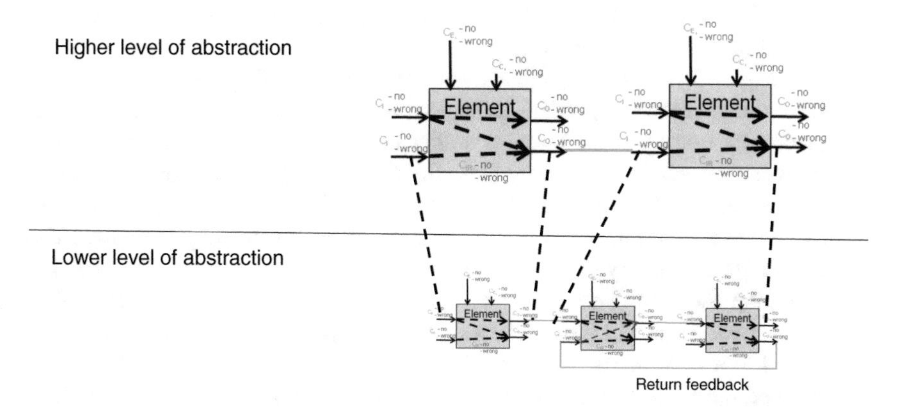

Fig. 4.44 Failure possibilities in a typical control system with return feedback e.g. close loop control

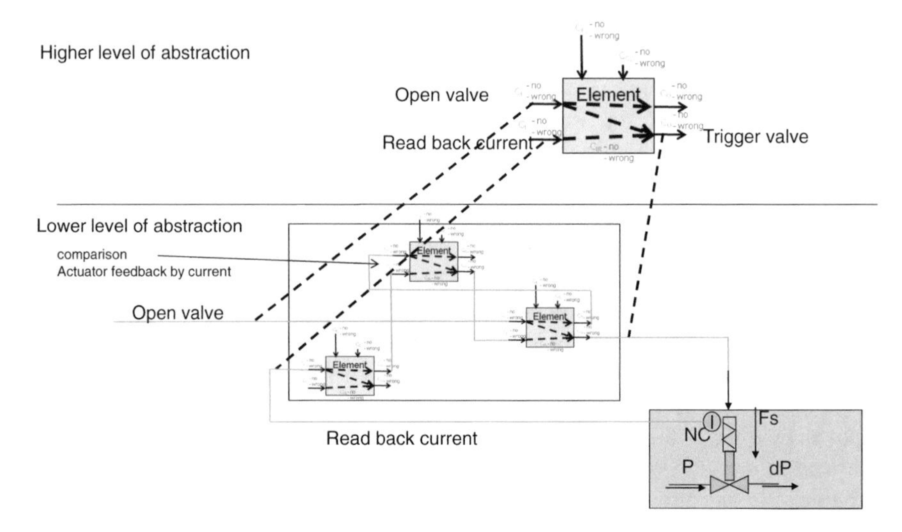

Fig. 4.45 Valve actuation with current monitoring

For a safe close-loop-controller there are more than just one requirement or more than one permitted operation mode, input condition, possible erroneous environmental influence factor, in many cases it is an array of modes and possible parameter. Safeguarding by a monitoring or by functional redundancy lead to the same complexity or heterogeneous implementation, than the close-loop-control itself. If the monitoring or the functional redundancy based on the same principles the same systematic errors lead to dependent failure and consequently the systematic-errors could not be controlled and consequently not avoided.

At the current read back of a valve current (see Fig. 4.45) stuck-at effects can already lead to higher complexity in the analysis. If the current to the valve is seen digitally, we will see whether the current, which is provided, is sufficient in order to open the valve. In this case we need to look at the physical environment of the valve and the corresponding spring force, which plays a role as counterforce for the electric design of the coil. Typical ageing effects, such as decreasing spring force, pressure or temperature dependent influences or sluggishness through the build-up of dirt can be compensated through sufficient tolerance limits. However, valve controls are often realized in a way that after the high current to move the valve a reduced current level are switched, that is sufficient to keep the valve open, but normally is insufficient to overcome the inertial torque for opening the valve. This has benefits for the energy or heat design of the valve and the control electronic and faster opening times can be possibly achieved. This could be relevant if the valves should open very fast, but also if the valve has to operate against a back-pressure, which lead to a high inertial force for the valve. The challenge is that there is no current fixed threshold for the set point, current temperature, aging effects or dirt at all involved parts could change the driving current for safe opening of the valve. The opening impulse has to be so high and so long that the valve opens safely.

Often we can see at the current or voltage profile (through the induction effect at the moving of the valve piston through the magnetic field of the coils) whether the valve opened. Then we can switch to the smaller holding current. However, the activation current can also be so high that currents are induced, which are too high and thus cable shielding, EMC safety measurements or other signal carrying elements are negatively influenced. If the valve for example closes undetected through vibrations the holding current still indicates an open valve, that the valve had been closed due to unexpected intensive vibration could not be detected. Furthermore, the question is whether a read back of the current can happen so quickly that the resolution of the input filter at an ADC of a microcontroller even permits such controlling strategy. With an analysis at an upper level of abstraction the entire failure analysis cannot be performed, since all these detail dependencies have to first be transparent in realization; all parts and their tolerances and the real aging effects in the real environment could be considered. If higher current-read-back or voltage monitoring provides a unique criterion could only conferment if the design verification is completed.

If such close-loop-controller or monitoring are implemented in a microcontroller, all internal failure of the microcontroller could lead to similar error impacts. Especially the memory effects, which can lead to a signal bouncing or to be stuck-at in certain conditions, are only analyzable at a very detailed level.

An alternative would be that these possible failures within the horizontal level are already intercepted in the level above so that in the case of failure, error propagations upwards to the safety goal could be avoided.

This can lead to a higher fault tolerance design, since entire error chains (errors in signal chains), which are implemented by means of higher level monitoring, could minor signal drifts or other short-time effects could compensated in the lower level signal chain itself (drifts too high in the coil could due to higher current compensated by inverse heat effects etc.). The higher level safety mechanism monitors only the resulting effect and only if the control-loop could not compensate itself, the monitoring should degrade the system.

Self-compensating control-loops in the lower (implementation) level provides stabile control conditions, so that higher level monitoring only detects a critical error degradation that could lead to violation of safety goals. In case of a well-design control loop it means only if the self-compensation mechanism fails itself. The specification for such monitoring further depends on the safety goals and the possible error modes, which can occur in the lower levels.

Error propagation in the verticals is often seen from the lower level in an upper level. However, systematic errors, for example from the determination of the safety goals, the differentiation of safety requirements to the components and electronic components or software units, are also errors, which can spread vertically. Specification errors are systematic errors, if a higher level safety requirement is wrong and a lower level safety requirement derives from a wrong higher level safety requirement, also vertical error propagation have to be considered. Without a verification of requirements through all level of abstraction down to the realization and sometimes down to production, errors could remain undetected until the integration tests in the descending branch of the v-cycle. If the integration tests systematically derives from the requirements in the descending branch of the V-cycle, than only intensive validations could detect the systematic errors.

4.4.2.3 Inductive Safety Analysis

The inductive safety analysis is described as a bottom-up method. It investigates unknown failure effects starting with known failure causes. Today the FMEA is the basic analysis method at all. It has been developed for almost twenty years in different ways. The classical form sheet analysis (blank table form analysis) can be called a truly inductive safety analysis, whereas the cause in this context is often also determined deductively. This means that potentially unknown causes are examined. All new FMEA methods start with the function, a task or characteristics of the basic parts and search for potential causes, which could lead to malfunction, wrong tasks or to deviations of required characteristics of the basic parts. The next step is the determination of error propagations so that the failure effect can be determined.

In ISO 26262 three types of FMEA are addressed, the System-, Design- and Process-FMEA. Depending on the standard the terms are referenced from the methods to differentiate the level of abstraction, the area of use or the methodology itself. In the new AIAG standards therefore mention the Design-FMEA at the system level and the Design-FMEA at the components level. For the verification of the functional safety concept we are still missing the safety analysis, which should give the answer, what safety mechanism controls what error modes. In this context the different company standards often refer to a **Draft-FMEA** or a **Concept-FMEA** at a very early design phase or during drafting the architecture. This is a method, which could usefully support the verification of the functional safety concept. The VDA standard describes generally a Product-FMEA that depending on the product focus or causes, failure of error level, is variably determined; the basic range based on the scope of the product In this case different set of measures are considered. It is distinguished between measures during development and during customer operation. Newer versions consider also a Mechatronic-FMEA, which considers more electronic hardware and specific analysis approaches. The Process-FMEA describes the analysis in the production process, whereas the Process-FMEA is often intertwined with the **Design-FMEA**. A malfunction (often the error level) could have its cause in the design (errors due to failure in the development) or in the production (errors due to failure in the production). In the sense of ISO 26262 the design of a product is practically examined through the Design-FMEA and corresponding measures are agreed upon in the development, which should reduce, mitigate or avoid the occurrence of failure or reduce, mitigate or avoid error propagations.

All requirements of ISO 26262 from part 5, Chap. 7 could be covered through the Design-FMEA, if the step of the function to the failure cause would be accepted as deductive analysis. The question now is if it is possible with a Design-FMEA to differentiation between the error or fault classification after single faults, multiple faults and safe faults. Single faults will be easily identifiable in the Design-FMEA because in this case error propagations could be identified into higher abstraction levels up to the safety goal. If a fault mode has the potential to propagate direct to a failure which could violate a safety goal, we call it a '**single-point fault**'. If a safety mechanism for this single-point fault exists, depending on the coverage of the fault mode the uncontrolled average called a 'residual fault'. If the fault leads to a safe failure or if a safe state could be achieved for a fault, without the potential to violate a safety goal, the fault can be identified as "**safe fault**". If there are no functional dependencies to safety relevant functions for a fault or an effect, which leads to the violation of a safety goals even in combination with other possible errors, such faults can be classified as non-safety relevant faults, unless the analysis of the dependent faults indicates again certain influences. For the multiple faults the norm points out further classifications such as the perception of the driver, those could also be technically detectable or latent, which means they could only violate safety goals if other malfunctions are present. A generic failure combination for multi-point-faults is already given. If a fault occurs in the function, even if it is a single fault, which could violate a safety goal and at the same time an implemented safety mechanism which has the task to control the fault could also have malfunctions, we have to consider already such failure combination, already as multiple-point-faults. **Multiple-point-faults**, which occur through the way of the realization, are often only identifiable by tests or simulations. Therefore, it is true that a fault tree analysis as a deductive analysis is a way to illustrate multiple-faults in their dependencies but in the context of a mere top down analysis failure combinations can only be derived by analysis functional dependencies also in case of faults or errors. Due to the rapid increasing number of possible failure combinations, only simulation could give answers to the possible or relevant failure combinations which lead to multiple-point-failure. Especially dependencies from systematic failure among themselves or systematic failure in combination with random hardware failure can only be derived through simulation and experience or by logical dependencies. Failure simulations, prototype tests with fault-injections and so on are possible "measures during the development" in the context of Design-FMEAs.

For the **Product-FMEA** according to the VDA-standard only different type of measures are distinguished, such as measure during development or during customer operation. Typical Design-FMEAs and specially the term System-FMEAs are not directly addressed. Due to the scope the Product-FMEA could be applied on vehicle, system, component, and in case of e.g. semi-conductors on sub-component (or part) level. How the structure and how the scope of an FMEA could be tailored based mainly on the complexity and on the product boundary (Fig. 4.46).

In the classical table based Form-Sheet-FMEA, it could be recognized, that we not performing a pure inductive or bottom-up analysis. We basic principle is to evaluate on a given function certain failure and in the following steps to identify failure causes and failure effects.

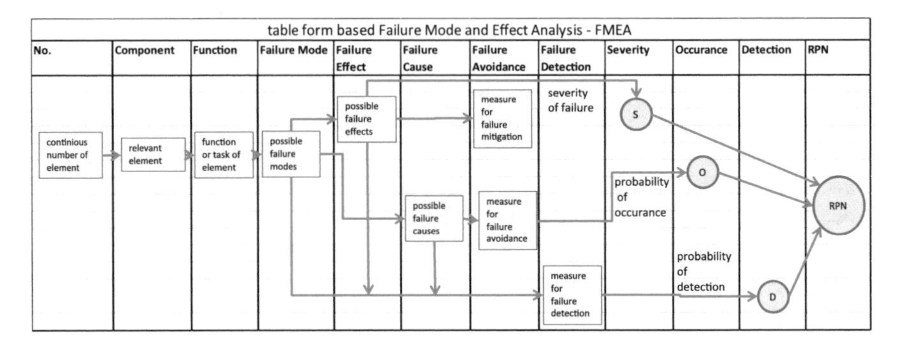

Fig. 4.46 Classical FMEA method (*Source* VDA FMEA 1996 [4])

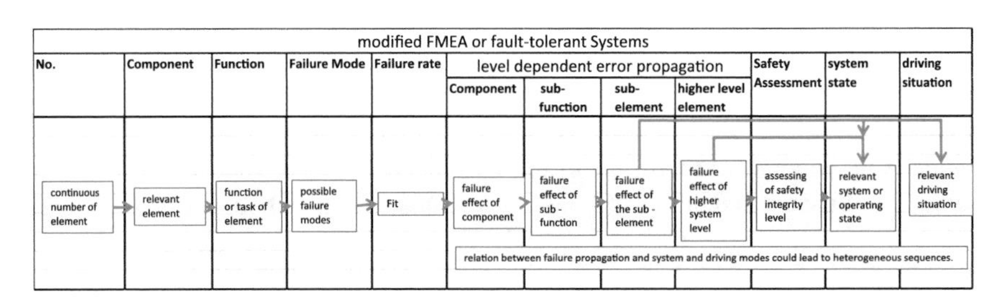

Fig. 4.47 FMEA for multiple system levels for control of multiple-point faults. (Translated Source: Marcus Abele [2], Modeling and assessment of highly reliable energy and vehicle electric system architecture for safety relevant consumers in vehicles, 2008)

If we arrange FMEA hierarchically beyond the classical sequence based on cause of failure, type of failure, failure effect, multiple system levels could be considered. In that different hierarchical system levels, error propagations could be avoided within the different level for example in the implementation, the component and system design, but also on vehicle level, for example between different ITEMs or vehicle systems (Fig. 4.47).

In general FMEA addresses only single-point failure. But also multi-point faults could be examined. This could be considered by the hierarchical structure so that the given safety architecture safeguards in a higher level the lower level malfunctions, errors, faults or failure could not lead to violations of safety goals. We have to consider than the lower level failure as multi-point faults, since the higher level mechanism could fail and then the lower level failure could lead to a safety goal violation. It is recommended, that malfunctions of the higher level safety mechanism could only lead to safe states like de-energizing, enabling of elements, resetting etc. If not the failure of the additionally implemented safety mechanism in the higher level become single-point failure. It means additional critical safety function would be implemented. In this case a further analysis needs to show that between the two functions no failure combinations lead to the violation of safety goals, which is required by ISO 26262 for ASIL C and D as an "Analysis of

Dependent Failure". These functional dependencies is an additional analysis, but also the usage of common resources on implementation level like microcontroller and on system level like common energy sources (battery, power supply etc.) have to be considered.

4.4.2.4 Deductive Safety Analysis

The deductive safety analysis is described as a top-down-approach. It examines unknown causes of failure starting with known failure effects. The old norm for the "Event Tree-Analysis", DIN 25424, did define the symbols which had to be used also as logical or Boolean elements for the Fault-Tree-Analysis. But for the analysis themselves, many different methodology had been developed.. Reliability block diagrams describe the logical dependencies in block diagrams and those block diagrams and their interfaces get analyzed. The result could even be represented in equations which uses Boolean algebra.

The quantifiable result in FTAs is often calculated as probability for the unavailability but it can also be considered and calculated positively as probability of availability.

The deductive analysis is in ISO 26262 required in additions to the inductive analysis for ASIL C and D elements. The aim in this case is to have a second independent analysis method, which analyzes the product independently from top-down and bottom-up. Therefore, the combination of the inductive analysis in one step with the deductive analysis is not well accepted. An automatic transformation of one analysis result into another illustration is also not expedient for safety. Therefore an alternative approach based on Reliability Block Diagrams could be considered.

The aim of the deductive analysis is primarily to detect possible failure before design decisions are made. Therefor the deductive analysis should be parallel during the developing and detailing of requirements, e.g. on the descending branch of the V-cycle.

The inductive analysis would consequently be the verification to see whether a design decision etc. is for example sufficient, appropriate, or adequate. This means that in the first iterations of the deductive analysis only information are available, which is derived from requirements, constraints etc. from higher abstraction levels. These could also be environmental conditions, systems or operation modes and architecture or design decisions or assumptions. This deductive analysis is required for ASIL C and D in the product development on system level (Part 4 of ISO 26262) and on hardware level (Part 5 of ISO 26262).

Since systems can be divided, structured or broken down hierarchically in subsystems, this analysis also needs to be applied in the respective sub-system or sub-components levels. If the systems and components are not hierarchical structured, the analysis themselves, as well their representation, become very complex. For error propagation an appropriate representation is often impossible, so that traceability could not be assured and never proper verified.

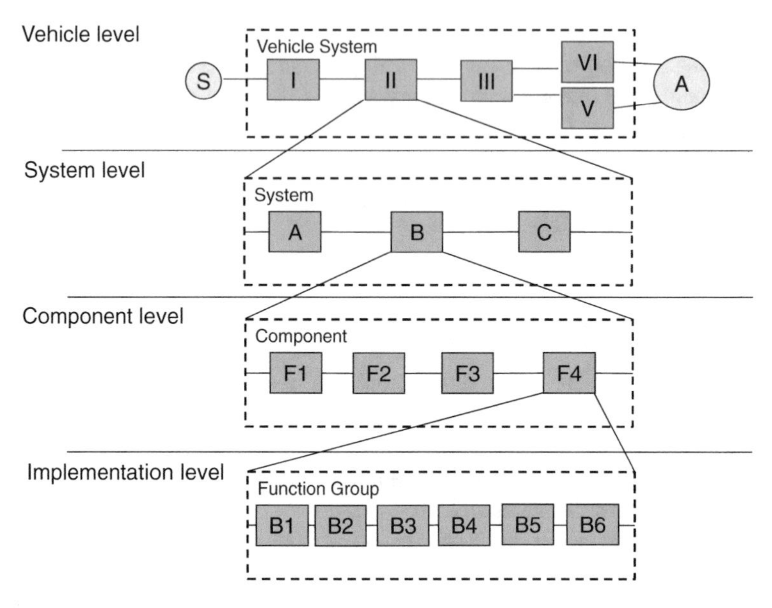

Fig. 4.48 Break-down of reliability block diagrams

Basically, for reliability block diagrams we also assume a systematical breaking down or decomposition of function blocks through the different abstraction levels. In the first iteration and on vehicle and system level logical function blocks are used, which are analyzed through the levels based on their functional dependencies. In the different levels signal chains could be extracted, which represent a continuous functional correlation and the required information-, or data-flow in the system. This could be applied for the intended function, but also the relation and the effects and influences of the evaluated safety mechanisms could be considered (Fig. 4.48).

At the lower realization level the basic electronic parts could be identified or even function groups inside semi-conductors. How deep the analysis have to be considers, dependents mainly on what level the lowest safety mechanism effects the relevant intended functions of the product.

The realization (see Fig. 4.49) at the lower level is portrayed in the following simplified circuit diagram.

This circuit diagram is transferred "inductively" in a logic chain (Fig. 4.50).

This logic chain can now be analyzed, whereas the aim of the analysis needs to be derived from the context or through the architecture up to safety goals. In this case, we would achieve a de-energized Safe-State for the coil B6 for the following single failure effects:

- high-resistance failure of the coil B6
- high-resistance failure of the resistor B2
- high-resistance failure of the redundant resistors B1 and B3
- high-resistance failure of the redundant transistors B4 and B5

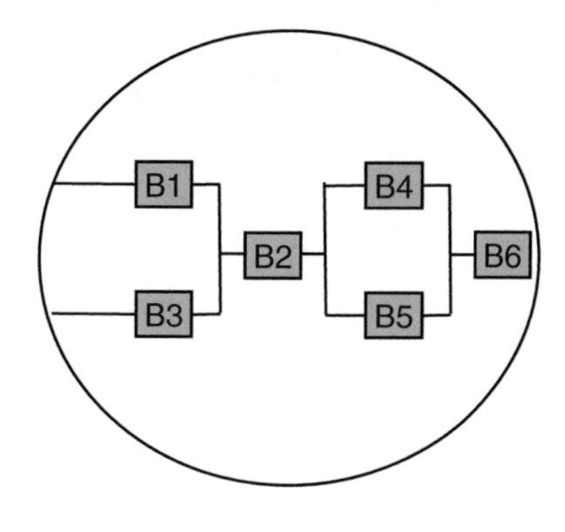

Fig. 4.49 Break-down of reliability block diagrams

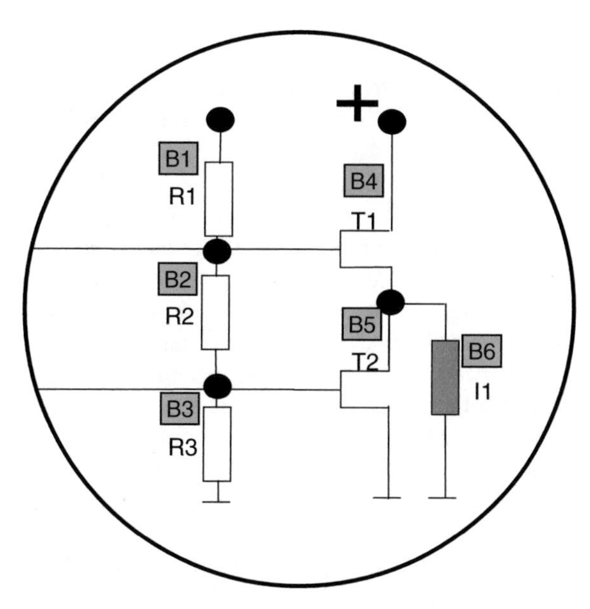

Fig. 4.50 Break-down of reliability block diagrams (typical driver design based on discrete HW parts)

If the derived safety requirement would now be the avoidance of an unintended switch on of the inductivity B6, certain failure modes could dangerously impact the functional groups in different intensities. All further low-resistance failures of the electric components would be seen as multiple-point faults. The following argumentation shows how important the correct design of the electric components can be even for such a simple circuit. Depending on the design, different failure modes can lead to single or multiple-point-faults.

R1 and R3 would be designed in a way that the transistors cannot both simultaneously be connected in case of a single-point-fault. Drifts on specific values of R2 can cause T1 and T2 to incorrectly switching. Also, a low-resistance fault of T1 could lead to an incorrect activation of the inductivity L1. With that the dangerous single-point faults are identified. In this context the safe faults as well as single- and multiple-point faults would be identified. With this design the low-resistance failure

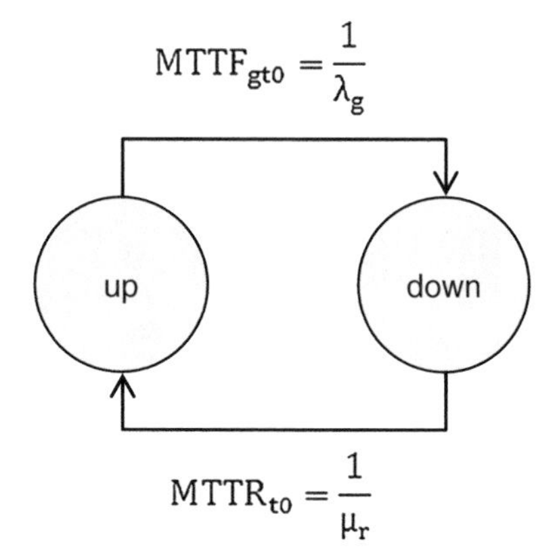

Fig. 4.51 Simple quantitative model for a repairable system

of T1 would still be a single-point fault despite the redundancy of the transistors. Wrong designed electronic in this case is a systematic error, the question is if a redesign of components for a higher ASIL would be the solution, or even a change of the general architecture, should be considered. The question of quantification would be only the second question.

The quantification of the reliability block diagrams is described precisely in the literature. There are mathematical approaches and the possibility to illustrate the result through Boolean algebra. The following basic elements are able to illustrate possible safety principals in the different structures. The mathematical derivations are described assuming a simple repairable system (Fig. 4.51).

The failure rate λ and μ are derived from the reciprocal of the MTTF (mean time to failure) or from the MTTR (mean time to restore). The index "g" means in this context "total", "t0" means starting from the $t = 0$ (time 0).

Therefore the basic functions are quantified as follows (Figs. 4.52, 4.53, 4.54 and 4.55).

Reliability block diagrams can be very well derived from the architecture and their functional correlations. Since it is relatively easy to transfer the function and its safety relevant characteristics through a simple mathematical correlation into the probability of default, the question is whether a negation even makes sense at the system level, at which the failures from systematic failures primarily show effect. What is the added value generated by a negation? Isn't it more important to check the identification of information and data flows through all the elements as well as the functional failure modes, which can dangerously impact the functions? Furthermore, we will be able to identify functional dependencies, since we can easily apply error-injection methods in the data nets or signal chains. It is not only possible to identify error propagations in vertical direction e.g. to the safety goals but also critical cascades, which lead to dependent failure within the horizontal levels, like faults in sensors that lead to wrong calculations in the microcontroller etc.

Identification of cascading failure addresses ISO 26262 as part of the Analysis of Dependent Failure, but even it is only required for ASIL C and D, cascading failure could also lead in ASIL A and ASIL B applications to violations of safety goals.

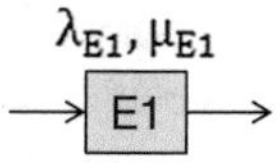

$$\lambda_{E1}, \mu_{E1} \quad \rightarrow \boxed{E1} \rightarrow \qquad PA_{E1} = \frac{1}{1+\frac{\lambda_{E1}}{\mu_{E1}}} \approx 1 - \frac{\lambda_{E1}}{\mu_{E1}}$$

Fig. 4.52 Probability of faults of a single element

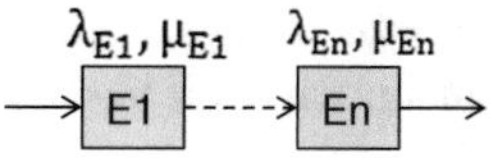

$$PA_g \approx 1 - \left(\frac{\lambda_{E1}}{\mu_{E1}} + \ldots + \frac{\lambda_{En}}{\mu_{En}}\right) \quad \lambda_g = \lambda_{E1} + \ldots + \lambda_{En} \Rightarrow \mu_g = \frac{\lambda_g}{1 - PA_g}$$

Fig. 4.53 Probability of fault of serial elements

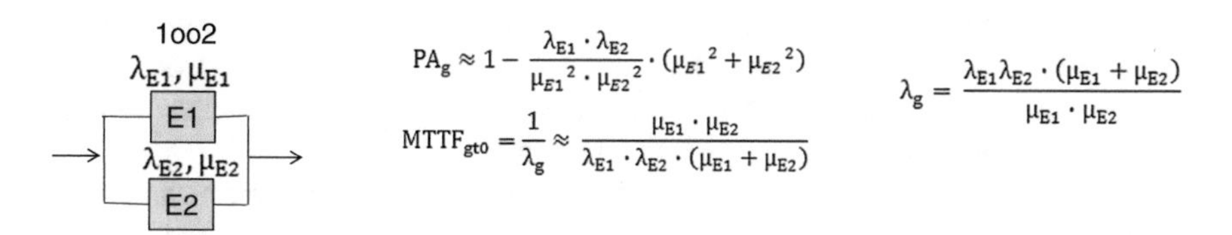

$$PA_g \approx 1 - \frac{\lambda_{E1} \cdot \lambda_{E2}}{\mu_{E1}^2 \cdot \mu_{E2}^2} \cdot (\mu_{E1}^2 + \mu_{E2}^2) \qquad \lambda_g = \frac{\lambda_{E1}\lambda_{E2} \cdot (\mu_{E1} + \mu_{E2})}{\mu_{E1} \cdot \mu_{E2}}$$

$$MTTF_{gto} = \frac{1}{\lambda_g} \approx \frac{\mu_{E1} \cdot \mu_{E2}}{\lambda_{E1} \cdot \lambda_{E2} \cdot (\mu_{E1} + \mu_{E2})}$$

Fig. 4.54 Probability of fault of parallel elements

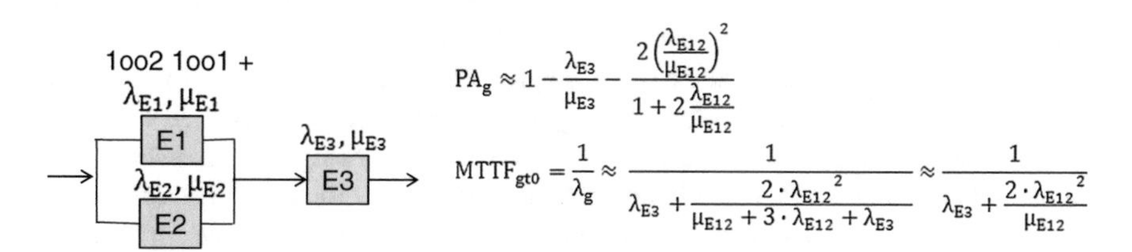

$$PA_g \approx 1 - \frac{\lambda_{E3}}{\mu_{E3}} - \frac{2\left(\frac{\lambda_{E12}}{\mu_{E12}}\right)^2}{1 + 2\frac{\lambda_{E12}}{\mu_{E12}}}$$

$$MTTF_{gto} = \frac{1}{\lambda_g} \approx \frac{1}{\lambda_{E3} + \frac{2 \cdot \lambda_{E12}^2}{\mu_{E12} + 3 \cdot \lambda_{E12} + \lambda_{E3}}} \approx \frac{1}{\lambda_{E3} + \frac{2 \cdot \lambda_{E12}^2}{\mu_{E12}}}$$

Fig. 4.55 Probability of faults of element combinations

Therefore we have the possibility to use the block diagrams for the functional dependency analysis. The dependencies through the technical realization are again to be considered by deductive analysis.

The analysis of the failure types is very essential (error or failure modes, possible error behavior of characteristics of elements etc.). ISO 26262 mentions indications in the correlating appendices of parts 5 (attachment D) and 6 (attachment D) for the safety mechanisms, which need to be implemented. For a deductive analysis we can only determine the possible failure modes from the function, the characteristics of the function (parameter) as well as their relation to the environment. Error modes like no function, an incorrect function; a function too low or too high or drifts can be evaluated in the context of their Diagnostic Coverage for electronic parts (DC). Furthermore, sporadic (intermittent or transient) failure, oscillations or other dynamic failure are derived from the specified intended functions and their characteristics. How and in what way these errors propagate, depends on environmental conditions. Thus, in a cold environment a failure can have a different impact than in a hot environment, for example when electric components are used at their specification limits. Therefore, a robust design is required in ISO 26262 or in other safety norms a so-called 'derating' (distance to the maximum or nominal design of electric components).

At least during the verification of the technical safety concept the result of the inductive and deductive analysis need to be merged. Which technical failures propagate further upwards to the safety goals in what way, how likely, and with which intensity, is then shown in the overall safety assessment, when all verification, integration and validation results are available.

However, the deductive analysis does not start at the point at which ISO 26262 first required it but already in the requirement analysis. Each Chap. 6 of the parts 4, 5 and 6 requires a verification of the requirements. Additionally, part 8 of Chap. 6 requires that the safety requirements are specified in natural language and in formal or semi-formal notations. Whereas according to the glossary of ISO 26262 the formal notation is a syntactically and semantically complete notation and the semi-formal notation is only a syntactically complete notation.

Semantic typically deals with the relations between signs and their meaning and the correlated statement; syntax defines the rules. Similar to a language, we can build sentences with the amount of provided symbols (words). The rules for the building of valid sentences from these symbols (="grammar rules") define the rules for the syntax. If for example we allocate a value to a variable or we use an inductive loop we need to respect the "grammar rules".

A wrong syntax leads to error messages for example during compiling of software. The meaning of valid sentences of a programming language is called semantic. It is all about the question what sign sequences cause in a computer: "2 + 4=7" is in the language of math syntactically correct but semantically incorrect. As a result the semi-formal method could provide despite the correct description wrong content results. At the first glance it is not clear why the formal notation isn't the preferred method. If a formal method is called upon a wrong context, it will provide wrong results for the wrong context also semantically and syntactically complete. Therefore, ISO 26262 mentions the semi-formal notation only as a possibility next to the natural language to formulate requirements. The sufficient completeness and correctness is determined by verification according to ISO 26262.

If we use the semi-formal notation to describe requirements, it is useful to also use this for the same basis of the model description. Because ISO 26262 requires a verification after each step, systematic failure could be avoided and the consistency of the work steps and therefore also the work results would be supported. Since the model is also used as test reference according to ISO 26262, the model matures alongside the development process, if the product model continuously validated versus the increasing maturity of the development samples or prototypes. A model is often based on logical elements or function groups. They describe the structure, functional correlations of the elements or their technical behavior accordingly. Therefore, the architecture, the safety analysis and the model should widely have a common basis or refereeing to the safety relevant characteristics at least, they should be consistent.

4.4.2.5 Quantitative Safety Analysis

There are two chapters in ISO 26262, Part 5, which cover the topic quantified safety analyses. The main referenced based on reliability analysis and failure rates for electrical parts, which addresses only random hardware faults. Therefore, the interpretation of the bathtub curve, the sufficient trust, or confidence in the used data and the significance of the determined results are always a question of how the analyses have been performed with which aim. Both metrics of ISO 26262 have different objectives. Part 5, Chap. 8 describes the following objective:

ISO 26262, Part 5, clause 8:

8.1 Objectives

8.1.1 The objective of this clause is to evaluate the hardware architecture of the item against the requirements for fault handling as represented by the hardware architectural metrics.

Although Part 5 addresses product development on hardware level this clause addresses the architecture of the entire Item, which means at least a complete vehicle system. It reduces also not only to the safety architecture, it references to the hardware architecture. By reading the "Objective" the question arises, what is the hardware architecture on system level? The answer gets another view if we look at the second part of the objective, which requires considering the defined architecture metrics to evaluate the architecture. The target of the architectural metrics seems to be that weak-points in the architecture should be controlled by adequate safety mechanism. Both the failure rate based fault mode and the percentage-based safety or control mechanism (DC for Diagnostic Coverage) and their efficiency should be defined in a quantitative measurable transparent metric.

ISO 26262, Part 5, Clause 8.2:

8.2 General

8.2.1 This clause describes two hardware architectural metrics for the evaluation of the effectiveness of the architecture of the item to cope with random hardware failures.

8.2.2 These metrics and associated target values apply to the whole hardware of the item and are complementary to the evaluation of safety goal violations due to random hardware failures described in Clause 9.

8.2.3 The random hardware failures addressed by these metrics are limited to some of the item's safety-related electrical and electronic hardware parts, namely those that can significantly contribute to the violation or the achievement of the safety goal, and to the single-point, residual and latent faults of those parts. For electromechanical hardware parts, only the electrical failure modes and failure rates are considered.

NOTE Hardware elements whose faults are multiple-point faults with a higher order than two can be omitted from the calculations unless they can be shown to be relevant in the technical safety concept.

8.2.4 The hardware architectural metrics can be applied iteratively during the hardware architectural design and the hardware detailed design
8.2.5 The hardware architectural metrics are dependent upon the whole hardware of the item. Compliance with the target figures prescribed for the hardware architectural metrics is achieved for each safety goal in which the item is involved.
8.2.6 These hardware architectural metrics are defined to achieve the following objectives:

- *be objectively assessable: metrics are verifiable and precise enough to differentiate between different architectures;*
- *support evaluation of the final design (the precise calculations are done with the detailed hardware design);*
- *make available ASIL dependent pass/fail criteria for the hardware architecture;*
- *reveal whether or not the coverage by the safety mechanisms, to prevent risk from single-point or residual faults in the hardware architecture, is sufficient (single-point fault metric);*
- *reveal whether or not the coverage by the safety mechanisms, to prevent risk from latent faults in the hardware architecture, is sufficient (latent-fault metric);*
- *address single-point faults, residual faults and latent faults;*
- *ensure robustness concerning uncertainty of hardware failures rates;*
- *be limited to safety-related elements; and*
- *support usage on different elements levels, e.g. target values can be assigned to suppliers' hardware elements.*

EXAMPLE To facilitate distributed developments, target values can be assigned to microcontrollers or ECUs.

Clause 8.2.2 considered to be complementary to the examination, which are required in Chap. 9. The main differences are not directly evident in the norm itself.
The Objective for the second clause is as follow:
ISO 26262, Part 5, Clause 9:

The objective of the requirements in this clause is to make available criteria that can be used in a rationale that the residual risk of a safety goal violation, due to random hardware failures of the item, is sufficiently low.

NOTE "Sufficiently low" means "comparable to residual risks on items already in use". Also the metric from clause 9 addresses the whole Item, which is again a whole vehicle system. Here is the focus not on the architecture of the system; the focus is more on a rational of the residual risk related to each safety goal.

The general requirement that derive from the objective addresses mainly the 2 possible methods to fulfil the requirements from clause 9.

Also these two methods are based on random hardware failure; they are defined in the Annex C of Part 5.

ISO 26262, part 5, appendix C:

The failure rate λ of each safety-related hardware element can therefore be split up as follows (assuming all failures are independent and follow the exponential distribution):

a) failure rate associated with hardware element single-point faults: λ_{SPF};
b) failure rate associated with hardware element residual faults: λ_{RF};
c) failure rate associated with hardware element multiple-point faults: λ_{MPF};
d) failure rate associated with hardware element perceived or detected multiple-point faults: $\lambda_{MPF\ DP}$;
e) failure rate associated with hardware element latent faults: $\lambda_{MPF\ L}$;
f) failure rate associated with hardware element safe faults: λ_S

then $\lambda = \lambda_{SPF} + \lambda_{RF} + \lambda_{MPF} + \lambda_S$ and $\lambda_{MPF} = \lambda_{MPF\,DP} + \lambda_{MPF\,L}$.

ISO 26262 presents in part 5, appendix C, Figs. C.2 and C.3, these correlations as pie chart (see Fig. 4.56) as well as mathematical formulas. The metrics are mentioned first in part 4 of ISO 26262 in Chap. 6 under the heading 'Avoidance of Latent Failure' (part 4, 6.4.4). In this context the requirement 6.4.4.3c demands the quantitative budgets for the top failure metrics. This demand is widely repeated in the requirement 7.4.4.3. The requirement 7.4.4.4 describes in such detail that for both metrics in part 5, Chap. 8 (Architecture Metrics) and part 9 (Top Failure Metrics), target values for the failure rates and diagnosis coverage should be specified.

Chapter 7, part 4 addresses the system design, the technical safety concept and their verification, which should be derived from the functional and technical safety requirements. Therefore, in requirement 7.4.3.1 the inductive (for all ASILs) and deductive (for the higher ASILs) safety analysis is required. In this context of product development on system level it is primarily a matter of the analysis of systematic failure. In one indication (note 1) it says that a quantitative analysis can support the results.

Consequently, no quantification is required in part 4, in order to analyze the system design. Only the planning of the targets for the metrics is required. Especially if we have to consider that "dependent failure" could be seriously impact

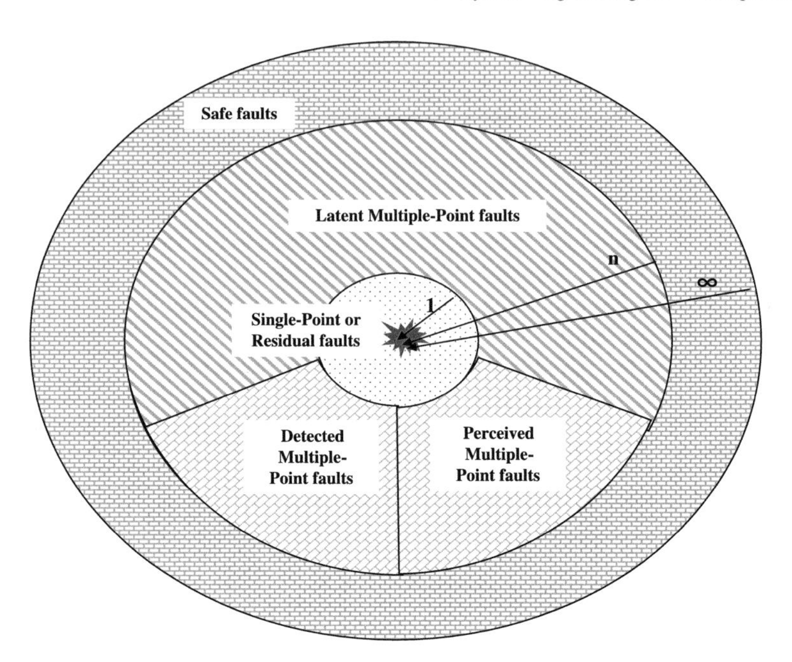

Fig. 4.56 Pie charts (Baumkuchen-Diagram) for the failure classification (*Source* ISO 26262, Annex C)

the correct functioning of the system, and leading to failure which could violate safety goals. In this context we detect for functional failures based on common causes or failure cascades that the essential resulting weaknesses in the design and in the realization. The challenge is to structure the failure analysis also in a way that the different dependencies of the components and the system environment can be considered or excluded; we have to define constraints and criterion for "sufficient independence" or "sufficient freedom from interference". In this case we will have no option but to actively define and specify separations mechanisms for the system architecture. This applies not only for the software, for example for the planning of a partitioning; there are also plenty of dependencies in the hardware, which cannot be considered in their full impact. For electronic hardware geometric distances, insolation etc. for parts, wiring, and harness and layout rules for the printed circuit board could be considered as solutions.

IEC 61508 published several models for the quantification of the dependency factor (Beta factor). However, it could not be included in ISO 26262, since it questioned the general validity of such quantification. The question is how to safely ensure a sufficient partitioning of the dependency especially of the functions and the signal line for the hardware, how could possible impacts be quantified? For the quantification according to ISO 26262 the single faults and the credible and possible failure combinations need to be excluded in any case. In the hardware design this is only possible by sufficient robust and fault tolerant design.

Consequently it is important to make sure that by developing the safety architecture separations should be considered and adequate barriers are specified so that also for all sub-elements quantitative budgets could be specified. It is generally known that failure can propagate according to all dependencies in the positive means along all functions, but in case of faults in complete other dependencies. In part 9, Chap. 7 (Analysis of Dependent Failure), the following list of possible impacts is provided by ISO 26262. In the following chapters of the book the list will be considered for higher level ASILs, but most of the criterion should be already considered by the general design of electronic hardware, even for QM functions.

4.4.2.6 Architecture Metrics

Target of the "Architecture Metrics" is making the safety architecture assessable and comparable.

The "Single-Point Fault Metric (SPFM)" is defined as follows:

ISO 26262, Part 5, Annex C:

C.2 Single-point fault metric
This metric reflects the robustness of the item to single-point and residual faults either by coverage from safety mechanisms or by design (primarily safe faults). A high single-point fault metric implies that the proportion of single-point faults and residual faults in the hardware of the item is low.
C.2.1 This requirement applies to ASIL (B), C, and D of the safety goal. The definition given by the following equation shall be used when calculating the single-point fault metric:

$$\text{Single-Point Fault metric} = 1 - \frac{\sum_{\text{safety-related HW elements}} (\lambda_{SPF} + \lambda_{RF})}{\sum_{\text{safety-related HW elements}} \lambda} = \frac{\sum_{\text{safety-related HW elements}} (\lambda_{MPF} + \lambda_{S})}{\sum_{\text{safety-related HW elements}} \lambda}$$

where $\sum_{\text{safety related HW elements}} \lambda_x$ *is the sum of* λ_x *of the safety-related hardware elements of the item to be considered for the metrics (Elements whose failures do not have the potential to contribute significantly to the violation of the safety goal are excluded from the calculations).*

In order to qualify elements or components or by evaluation a DC by means of Monte-Carlo-Simulations, the following formula could be considered:
ISO 26262, Part 5, Annex C:

The failure rate assigned to residual faults can be determined using the diagnostic coverage of safety mechanisms that avoid single-point faults of the hardware element. The following equation gives a conservative estimation of the failure rate associated with the residual faults:

$$DC_{\text{with respect to residual faults}} : \text{Diagnostic Coverage as a percentage}$$

$$DC_{\text{with respect to residual faults}} = \left(1 - \frac{\lambda_{\text{RF estimated}}}{\lambda}\right) \times 100$$

$$\lambda_{RF} \leq \lambda_{\text{RF estimated}} = \lambda \cdot \left(1 - \frac{DC_{\text{with respect to residual faults}}}{100}\right)$$

The "Latent-Fault-Metric (LFM)" is defined as follows:
ISO 26262, Part 5, Annex C:

C.3 Latent-fault metric
This metric reflects the robustness of the item to latent faults either by coverage of faults in safety mechanisms or by the driver recognizing that the fault exists before the violation of the safety goal, or by design (primarily safe faults). A high latent-fault metric implies that the proportion of latent faults in the hardware is low.
C3.1 This requirement applies to ASIL (B), (C), and D of the safety goal. The definition given by the following equation shall be used when calculating the latent-fault metric:

$$\text{Latent Fault metric} = 1 - \frac{\sum_{\text{safety-related HW elements}} (\lambda_{\text{MPF Latent}})}{\sum_{\text{safety-related HW elements}} (\lambda - \lambda_{SPF} - \lambda_{RF})}$$
$$= \frac{\sum_{\text{safety-related HW elements}} (\lambda_{\text{MPF perceived or detected}} + \lambda_S)}{\sum_{\text{safety-related HW elements}} (\lambda - \lambda_{SPF} - \lambda_{RF})}$$

where $\sum_{\text{safety-related HW elements}} \lambda_x$ is the sum of λ_x of the safety-related hardware elements of the item to be considered for the metrics (Elements whose failures will not have the potential to contribute significantly to the violation of the safety goal are excluded from the calculations).

In order to determine DC the formula could be consider as follow:

The failure rate assigned to latent faults can be determined using the diagnostic coverage of safety mechanisms that avoid latent faults of the hardware element. The following equation gives a conservative estimation of the failure rate associated with latent faults:

$$DC_{\text{with respect to latent faults}} : \text{Diagnostic Coverage as a percentage}$$

$$DC_{\text{with respect to latent faults}} = \left(1 - \frac{\lambda_{\text{MPF L estimated}}}{\lambda}\right) \times 100$$

$$\lambda_{\text{MPF L}} \leq \lambda_{\text{MPF L estimated}} = \lambda \cdot \left(1 - \frac{DC_{\text{with respect to latent faults}}}{100}\right)$$

NOTE 2 For this purpose, Annex D can be used as a starting point for diagnostic coverage (DC) with the claimed DC supported by a proper rationale.
NOTE 3 If the above estimations are considered too conservative, then a detailed analysis of the failure modes of the hardware element can classify each failure mode into one of the fault classes (single-point faults, residual faults, latent, detected or perceived multiple-point faults or safe faults) with respect to the specified safety goal and determine the failure rates apportioned to the failure modes. Annex B describes a flow diagram that can be used to make the fault classification.

These architecture metrics are based on reliability data of elements and bring them in relation to the implemented control mechanism, which are the implemented safety mechanism. Random hardware failures of the electronic components are used as a basis for the data.

In general the architectural metrics could be considered as a process metric. The following activities are necessary to fulfill the metric requirements:

- Identification of all elements or electrical parts involved in the safety-related function (identification of non safety-relevant elements also called don't care elements or parts in other standards).
- Identification of the safety-related signal chain (Safety-related information or data flow e.g. to the microcontroller (Logic Solver) and the actuator and vis verse).
- Identification of elements or parts which are relevant for the functions related to the specific safety goals.
- Identification within the boundary analyzed before all "safe" elements or parts, that mean, elements that could fail however, could not violate safety goals even in combination at least in second order (Multiple-Point-Faults).

- For all residual elements or parts, which have somehow the potential to violate given safety goals, it should be identified, if their fault-modes could propagate direct to a safety goal violation, than these fault-modes have to be considered as single-point faults, if only indirect or by order higher than 2 they are multiple-point fault.
- Identification of already implemented redundancies or monitoring, which could be considered as a safety mechanism.
- Identification or evaluation of all required safety mechanism and their effectiveness by using tables in Part 5, Annex D or by other methods like Monte-Carlo-Simulations etc.
- Optimizations, unless the targets for the metrics are fulfilled.
- Specification of all implemented (or to be implemented) safety mechanism based on the analysis.
- Development of a test concept to show sufficient effectiveness of all safety mechanism.

Through the quantification the failure probabilities and the effectiveness of each safety mechanisms become comparable and an assessable. It is not clearly stated in ISO 26262 at which level the lower limits for the assessment need to be set or the process chains should run. An element is referenced for the metrics, which means it is not general necessary to trace down on electrical part level. Since ASIL B is in brackets, a higher element level could be considered, such as functional blocks and the main arguments for the metrics derived from the architecture and safety mechanism like current read-back from the actuator etc. The basis data can be used from known table out of data manuals like reliability hand-books etc., field data or by expert judgment. However, in Chap. 8 we do not find a reference to part 5, appendix F, since the precise quantification would not be expedient in this case.

The focus for the data evaluation for the architectural metrics is defined as follow:

ISO 26262, Part 5, clause 8.4.7:

8.4.7 This requirement applies to ASIL (B), C, and D of the safety goal. For each safety goal, the whole hardware of the item shall comply with one of the following alternatives:

a) to meet the target "single-point fault metric" value, as described in 8.4.5, or
b) to meet the appropriate targets prescribed at the hardware element level which are sufficient to comply with the single-point fault metric's target value assigned to the whole hardware of the item, given in requirement 8.4.5, with the rationale for compliance with these targets at the hardware element level.

NOTE 1 If an item contains different kinds of hardware elements with significantly different failure rate levels, the risk exists that compliance with the hardware architectural metrics only focus on the kind of hardware elements with the highest magnitude of failure rates. (One example where this can occur is for the single-point fault metric for which compliance can be achieved by considering the failure rates for failures of wires/fuses/ connectors, while disregarding the failure rates of hardware parts with significantly lower failure rates). The prescription of appropriate metric target values for each kind of hardware helps to avoid this side effect.

NOTE 2 The transient faults are considered when shown to be relevant due, for instance, to the technology used. They can be addressed either by specifying and verifying a dedicated target "single-point fault metric" value to them (as explained in NOTE 1) or by a qualitative rationale based on the verification of the effectiveness of the internal safety mechanisms implemented to cover these transient faults.

NOTE 3 If the target is not met, the rationale for how the safety goal is achieved will be assessed as given in 4.1.

NOTE 4 Some or all of the applicable safety goals can be considered together for the determination of the single-point fault metric; but in this case the metric's target to be considered is that of the safety goal with the highest ASIL.

Note 1 provides the target for the data on consistency rather than precision of the quantified data.

The most important result of a quantified analysis is more the average and related fault-modes which is "undetected" rather than looking at the result of the metric calculation and the result themselves. Therefore, a distribution of the failure modes of the electronic components in detail is not even that important. This is why it makes no sense to really use other failure distributions for the architecture metrics than those Alexandre Birolini published in his book. For a failure average of an electronic component of less than 10 % an assessor will possibly become skeptical and could check, which influence a higher value could have on the result. Of course, for example short-circuit-proof capacitors exist, but in this case we could also bring credible arguments. The effectiveness of the safety mechanism based on analogies to tables in part 5, appendix D. Diagnosis coverage (DC) significant less than 90 % will not be questions at all, because if there is any safety mechanism at least half of the fault-mode (50 %) could be always covered. But if a safety mechanism could cover within the entire specification space all the fault-modes with 99 % or even more, could not be easily shown. It is useful to verify the effectiveness of all diagnoses (DCs) by appropriate fault-injection tests.

The target value for the architecture assessment could also be derived from a comparable design. However, it is questionable, if all information are available from the comparable design, and if the relevant environment and all the relevant functions, and technical impacts are really the same. Even the additional overhead needed to prove that these 2 designs and architectures are really the same or sufficient equivalent, could mean an enormous effort.

ISO 26262 only recommends the architecture metrics for ASIL B and requires the latent failure metric only for ASIL D functions. If the quantitative approach from IEC 61508, often called FMEDA(forms mainly in MS-Excel and based on part-count method with the fault-distribution as described by Alexandre Birolini, see also ISO 26262, Part 5, Annex E), is recommended or a more deductive approach should be applied could depend on the application. A deductive approach could provide also insights related to systematic faults and non-functional failure, by a pure part-could approach, this could be questioned.

Normatively are only the fulfillment of requirements and the normative results of the metrics required. In order to support a verification of the electronic design an inductive quantitative analysis, which considers the error propagation to the safety goal would be recommended, but would it be the target of the architectural metrics or more for the metrics required in Part 5, clause 9 of ISO 262626? The causes of failure can be determined deductively at the electronic components in a functional electronic group; such groups could be considered as electronic elements. Through the qualitative failure propagation of the relevant failures to the highest malfunctions (malfunctions that could violate safety goals), quantifications could be assessed and classified through a calculations or a Monte-Carlo simulation. Consequently, it would not be considered a pure inductive or deductive analysis. It could be a deductive analysis from functional error modes of electronic or functional groups up to safety goals, and for the critical elements an inductive analysis related to all faults modes of relevant hardware parts. This has the advantage that in the consideration also the functional electronic groups are tested according to their sufficient robust design for example accompanied by a Design-FMEA and therefore the stress factors (e.g. pi factors) for the failure rates are could be tested as part of the design verification process, so that the requirements from part 5, Chap. 7 are widely fulfilled. Without a sufficiently robust design, which is required in Chap. 7, safety architecture also isn't sufficiently robust itself. Furthermore, there are advantages to see that the tables in appendix D and the suggestions for the quantifications of the diagnosis coverage also cover systematic errors in the electronic components and their environment.

If a resistor is safe-guarded against open circuits, the printed circuit board, the junctions and the solder joint of the resistor are also safeguarded against possible open circuits. Especially safety mechanisms, which work at higher levels, for example on system level, could control also systematic faults in entire signal chains (e.g. from sensor to the software interface in a microcontroller. Safety mechanism on higher system level could lead also to higher availability and/or better failure tolerance. Such an implemented safety mechanism could only react to on critical failure behavior of electronic components any non-critical error could be tolerated or the threshold of the diagnosis could be adjusted exactly to the critical level.

If an ASIL A (ASIL A(D)) function is part of an ASIL-Decomposition of ASIL D function, the signal chain also for the implementation in ASIL A have to be quantified. If all possible failures of an ASIL B function have a safety mechanism with diagnosis coverages higher than 90 %, it can also be argued that through the percentage of safe failure the target goals of more than 90 % can also be achieved without a quantification of the detailed fault modes of the hardware parts. Often also the architecture metrics are used as an abort criterion, since part 4 requires considering corresponding safety mechanisms for all possible systematic failures. In case of using current or voltage read-back for safety-relevant actuator functions and functional redundancy for sensing of safety-relevant effects, to prevent systematic failure, also the random hardware faults are sufficiently covered at least by a SPFM of better than 90 %, which is sufficient for ASIL B.

If safety mechanisms against all possible systematic failures would be implemented at the system level, all random failures in the E/E hardware are also covered. By adequate verifications and integration according ISO 26262 any further design error in the components could be identified.

Main analysis for the architectural metrics is to make the safety relevant signal chains transparent and add adequate safety mechanism in case of weaknesses. Inspired by Robert Lusser, the signal chains are a chain of elements and the weakest parts should be enforced by means of safety mechanism. A typical safety mechanism consist of a part that can detect, malfunctions such as fault, errors or failure and a part that could control the malfunctions. It should be able to degrade the system to a safe state or switch to dissimilar redundant functions, which are identified as error free during runtime. Therefore, the entire signal chains and its elements (chain links) need to be identified. The quantification after Erich Pieruschka is primarily used to make the strengths of the chain links comparable. What is important: The safety relevant function is first subject of the analysis. The correct functioning of the safety relevant function has to be assured. If this is provided by adequate measures such as implemented safety mechanism and control measures, this forms architecture to safety architecture.

The quality of the detection and the level of control of fault or error modes are quantified as diagnosis coverage (DC). With help of this quantification the entire safety architecture could be quantified so that the degree of safety, effectiveness related to safety becomes comparable, measurable and assessable.

The identified weak chain links of the safety architecture are then the essential input for top failure metrics. The weak links now need to be assessed based on the realized design, which should be task of the following metric.

4.4.2.7 Top Failure Metrics (Probabilistic Metric for Random Hardware Failure, PMHF)

ISO 26262 describes two alternative methods to assess the influence of failures in the design or realization in relation to the safety goals. The first method considers a quantitative evaluation of the probability that random hardware faults violate a specific safety goal. Alternatively it is assumed that in a safe design and its correct realization, about one hundred single-point or residual faults could be identified, which have the potential to violate a specific safety goal. This is why per single-point, residual or possible failure combination 1 % of the target values is determined for each single or remaining fault in the safety-relevant system. This is an interesting approach for a component development, for which the system integration is unknown. It generally leads to very conservative analyses. If the cutsets at the higher system levels are analyzed in more detail, the errors propagate with far less probability to the safety goal. However, the method could be an interesting approach, since the metric requires also a very reliable system. This method banks on the fact that the occurrence of faults can be avoided or its probability drastically reduces.

The top failure metrics officially called PMHF (Probabilistic Metric for random Hardware Failures) in ISO 26262. It represents a comparable metric such as PFH (Probabilistic Failure per Hour) of IEC 61508. The top failure metrics of ISO 26262 focuses on failure probabilities, with which a safety goal could be violated, whereas PFH according to IEC 61508 is all about the probability of a danger through the system. Both target values of the metrics are specified in failure per hour (failure in time, FIT = 10E−9 h). Also in this case we assume an exponential distribution of the basis failure rate. The key difference between PFH and PMHF is that the PMHF is per safety goal and PFH for a safety-related system. The PFH considers mainly the probability that the system reaches in case of failure a de-energized safe state.

According to ISO 26262 there are three different alternatives for the quantitative goals.

ISO 26262, Part 5, Clause 9.4.2.1:

9.4.2.1 This requirement applies to ASIL (B), C, and D of the safety goal. Quantitative target values for the maximum probability of the violation of each safety goal due to random hardware failures as required in ISO 26262-4:—, 7.4.4.3 shall be defined using one of the sources (a), (b) or (c) of reference target values as outlined below.

NOTE 1 These quantitative target values derived from sources (a), (b), or (c) do not have any absolute significance and are only useful to compare a new design with existing ones. They are intended to make available design guidance as de-scribed in 9.1 and to make available evidence that the design complies with the safety goals.

a) Derived from Table 6, or
b) Derived from field data from similar well-trusted design principles, or
c) Derived from quantitative analysis techniques applied to similar well-trusted design principles using failure rates in accordance with 8.4.3.

NOTE 2 Two similar designs have similar functionalities and similar safety goals with the same assigned ASIL.

Since ISO 26262 is still new in the automotive industry, it will currently be difficult to get the derivation of the target value from field data or statistical calculation methods. ASIL C and D systems haven't been around and unchanged for long with the same operating conditions. It would be very ambitious to come up with any statistical hypothesis to the quantification without such a field experience. This is why in practice often only the Table 6 is considered (Fig. 4.57).

Since this metric follows in the standard one clause later than the architecture metrics, the focus is more on the realized design and not on the architecture. This is why it is questionable whether the same values for the random hardware failures as for the architecture metrics can be used. If the EE hardware of an ITEM really has 100 minimal cutsets or one hundred single-point or residual faults, controlled single faults (remaining fault percentage) or credible error combinations with an order higher than 2, it will be very hard work for the design to prove it. The identification of all safety-relevant first-order cutsets at least is formally given by the architecture metrics and the analysis of the dependent failures. Quantification is often difficult, since the realization for example relies for all environmental impact on the robustness of the design, rather than on random hardware faults. Systematic errors in semiconductors, **electromagnetic immunity (EMI)** or their **electromagnetic compatibility (EMC)** heat dependent errors could also lead to violation of safety goals, but the quantification and their relation to random hardware errors are not quantifiable, since the relation depends on to many factors. In many cases only sufficient robustness, conservative design and expert judgement e.g. by analogies to similar cases could provide safety arguments. Complementary statistical stress tests lead only to results, if the number of influencing stresses is limited.

The second alternative method considers being very conservative, but it does not support to identify impacts of systematic errors on the error propagation. Therefor the method could not provide further safety arguments. Some of the safety-relevant failure could be identified by the Analysis of Dependent Failure, but the analysis and the metric don´t any systematic approach about the probability of error propagation and possible or probable potential to violate safety goals. System with multiple safety goals, for which the error propagation to the safety goal is already very heterogeneous, due to overlapping of failure modes between safety goals such structure lead even to more combinations so that only qualitative arguments could be provided.

ASIL	Random hardware failure target values
D	$< 10^{-8}\ h^{-1}$
C	$< 10^{-7}\ h^{-1}$
B	$< 10^{-7}\ h^{-1}$
NOTE These quantitative target values described in this table can be tailored as given in 4.1 to fit specific uses of the item (for instance if the item is able to violate the safety goal for durations longer than the typical use of a passenger car).	

Fig. 4.57 Target values for top failure metrics (*Source* ISO 26262 part 5, Table 6)

The minimal cutsets for systems are not only on E-hardware level, they are more on system level. The probability of the error propagation on system level is rather determined by the systematic error influences than the quantitative probability of the occurrence of random hardware failure. This analysis is often also called sensitivity analysis or importance analysis (Fussell-Vesely importance, Birnbaum importance etc.). Through the analysis and definition of the relative influence of individual basis events at the default probability of a top event, quantification is also possible. Whether the result of such an analysis of the importance is actually represented in the form of a tree or more clearly arranged in form of spreadsheets, should result from the concrete analysis. In this analysis we don't determine the position in the hierarchy, which should already be considered through the architecture metrics. Furthermore, in Chap. 9 we no longer discuss whether we analyze inductively or deductively. At this point the focus lies on the assessment of the cuts in the system. Caution should be exercised in this context for the failure combinations of systematic and random hardware failure. Especially design related failure such as signal cross talking; EMC or heat influences essentially change the importance and therefore the probability to propagate to safety goals. These influences are often very difficult to quantify. For an analysis of dependent failure ISO 26262 doesn't explicitly require such an analysis for the lower ASIL as well as for the cutsets where no functional redundancies occur.

The architecture metrics are primarily used to assess the architecture. Top failure metrics rely on the realized design—the final product. Therefore, there are essential more in depth requirements for the accuracy of the failure rates. Their influence factors and the relation of the results are often based on different data sources.

ISO 26262, suggests in part 5, appendix F the following recalculation for the different data sources for the top failure rate and causes of failure:

ISO 26262, Part 5, Annex F:

Therefore, in the calculations, different failure rates sources can be used for different hardware parts of the item. Let Ta, Tb, and Tc be the three possible sources for the definition of the target values for the PMHF and Fa, Fb, and Fc be the three possible sources for the estimation of a hardware part failure rate. Let $\pi_{Fi \to Fj}$ be the scaling factor between sources Fi and Fj. This factor can be used to scale a hardware part failure rate based on Fi to a failure rate based on Fj.

$$\pi_{Fi \to Fj} \text{ can be defined as } \pi_{Fi \to Fj} = \lambda_{k.Fj} / \lambda_{k.Fi}$$

Where

$\lambda_{k.Fj}$ is the failure rate for a hardware part using Fj as the source for the failure rate; and
$\lambda_{k.Fi}$ is the failure rate for the same hardware part using Fj as the source for the failure rate.

In this case, knowing the corresponding scaling factor enables the scaling of a similar hardware part failure rate based on Fi to a failure rate based on Fj:

$$\lambda_{l,Fj} = \pi_{Fi \to Fj} \times \lambda_{l,Fi}.$$

The following chart provides an overview and the relation of the different Pi factors.

ISO 26262, Part 5, Annex F:

Table F.1 shows the possible combinations of target values and failure rates
NOTE 1 The targets of Table 6 are based on calculations using handbook data and under the assumption that handbook data are very pessimistic.
NOTE 2 If the source of data for the target and for the hardware part failure rate are similar, then no scaling is necessary.

Table 1 Possible combinations of sources of target values and failure rates to produce consistent failure rates for use in calculations

		Data source for Target Value		
		Table 6 9.4.2.1 a	Field data 9.4.2.1 b	Quantitative analysis 9.4.2.1 c
Data source for failure rates of hardware parts	Std. Database 8.4.3 a	$\lambda_{k,Fa}$ (1)	$\lambda_{k,Fb} = \pi_{Fa \to Fb} \times \lambda_{k,Fa}$	(2)
	Statistics 8.4.3 b	$\lambda_{k,Fa} = \pi_{Fb \to Fa} \times \lambda_{k,Fb}$	$\lambda_{k,Fb}$	(2)
	Expert judgment 8.4.3 c	$\lambda_{k,Fa} = \pi_{Fc \to Fa} \times \lambda_{k,Fc}$	$\lambda_{k,Fb} = \pi_{Fc \to Fb} \times \lambda_{k,Fc}$	(2)

(1) For some types of hardware parts, different handbooks can give different estimates of the failure rate of the same type of hardware part. Therefore the scaling factor can be used to scale the failure rates of a hardware part using different handbooks

(2) To have a consistent approach, failure rates have the same origin as the failure rates used in the calculation of the target value

The table considers target values on vehicle level also in relation with data for hardware parts. Getting target value for a new function from field data would be very questionable and as all ready discussed of comparable systems and their quantification are really available is also very doubtful. Therefor it seems to be very probable to use the data from Table 6. ISO 26262 provides 2 examples for the recalculation:

ISO 26262, Part 5, Annex F, Example 1:

EXAMPLE 1 Evidence can be made available that $10^{-8}/\mathrm{h}$ *with a 99 % level of confidence is similar to* $10^{-9}/\mathrm{h}$ *with a 70 % level of confidence. Therefore failure rates based on a recognized industry source considered with a 99 % level of confidence can be scaled to failure rates based on statistics with a 70 % level of confidence using the scaling factor* $\pi_{Fa \to Fb} = \frac{10^{-9}/\mathrm{h}}{10^{-8}/\mathrm{h}} = \frac{1}{10}$ *or the other way round.NOTE 3 Based on experience, a 99 % level of confidence can be considered for failure rates based on recognized industry sources as referred to in 8.4.3.*

ISO 26262, Part 5, Annex F, Example 2:

EXAMPLE 2 From a previous design, calculated failure rates from a data handbook and warranty data have been obtained. We know that

$$\lambda_{handbook} / \lambda_{warranty} = \pi_{Fb \to Fa} = 10.$$

where

$\lambda_{handbook}$ *is the calculated failure rates from a data handbook,*
$\lambda_{warranty}$ *is the calculated failure rates from warranty data and*
$\pi_{Fb \to Fa}$ *is the resulting scaling factor.*

If in a new design, we use the handbook data to determine the failure rates except for one hardware part (hardware part 1) for which we have only warranty data, then we can determine the handbook scaled data for this hardware part, $\lambda_{1,handbook} = \pi_{Fb \to Fa} * \lambda_{1,warranty}$.
Where

$\lambda_{1,handbook}$ *is the failure rate of the hardware part 1 using handbook data and*
$\lambda_{1,warranty}$ *is the failure rate of the hardware part 1 using warranty data.*

For instance, if $\lambda_{1,warranty} = 9 \times 10^{-9}/\mathrm{h}$, *then* $\lambda_{1,handbook}$ *can be calculated as* $9 \times 10^{-9} * 10 = 9 \times 10^{-8}/\mathrm{h}$.
Using this $\lambda_{1,handbook}$, *a consistent evaluation of the violation of the safety goal due to random hardware failures can be done.*

For practical use, these examples propose for the top failure value data from Table 6 and data from field observation could be scaled by a factor 10 with data from handbooks. Especially for the real stress which effected the hardware parts from field observation is not anymore traceable, as a consequence the factor 10 seems to be sufficient conservative estimation for the data in relation to handbook data which give guidelines how to deal with stress factors for heat, voltage, current etc.

4.4.2.8 Failure Metrics for Sensors or other Components

All metrics are based on an item, which are at least a vehicle system and the respective safety goals. How a single sensor or another components could be quantified, which could be also integrated in many different ways, is not really considered in ISO 26262. The question arises, what are the target values and what are the typical stress figures for base failure rate? For the architecture metrics a single channel system based on ASIL D components a DC_{SPF} of 99 % has to be achieved. Whether this value can be achieved only with measures within the component limits or also with external measures is a difficult question.

Measures or implemented safety mechanisms cost money, resources and development time, which are always difficult to be aware of, if it is not planned ahead of the development of an entire system. It is even more difficult if such components are run in ASIL decomposition. In this case there may be three parties involved, which have to come to an agreement for the failure control, the redundant parts and most likely a common element such as voter, comparator or similar.

Two diverse sensors and a separate electronic control unit would be an often-realized ASIL decomposition. However, it is not all about distributing the measures to the elements involved, but it is necessary to figure out, which measures or safety mechanism are even necessary against which failures. In order to do so, we would need the failure analysis of both sensor signal chains and details of the possible failure effects at the interfaces of the elements. If the safety mechanisms should be implemented in the electronic control unit, the specification of the failure effects of the redundant sensor signal chain builds the foundation of those safety mechanisms. The main safety effect based on the fact that it is unlikely that a certain failure effect occurs simultaneously in the redundantly implemented signal chain. If the errors in the signal chain do not occur at the same time a comparator could detect unequal information as an output of the signal chain. To quantify and specify such failure effects and a deterministic prognosis for example in which operating or driving situation they occur could be challenging. Without detailed behavior of the failure effects it cannot be evaluated whether the failure can be safely and reliably detected by the comparator. The advantage of such an approach is that the comparator can be set in a way that it actually only switches off when the occurring failure would otherwise propagate to a safety goal.

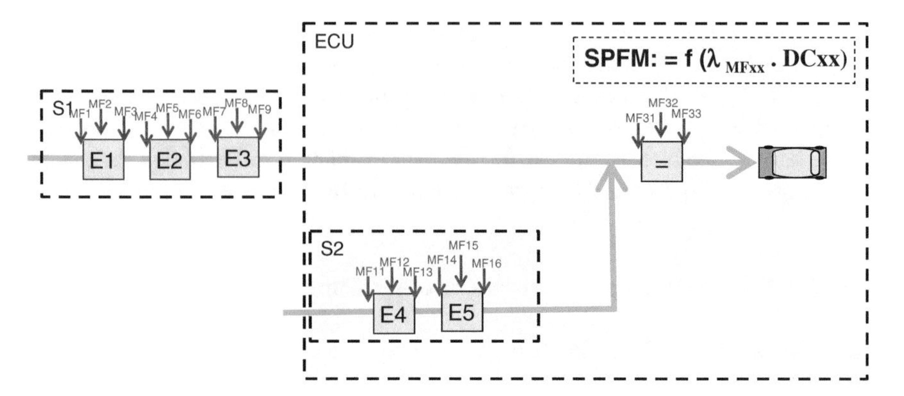

Fig. 4.58 Errors in redundant signal chains for a sensor inside and outside of the boundary

If an ASIL decomposition (see Fig. 4. 58) would consist of these two sensor chains (S1 and S2) as well as the electronic control unit (ECU), all errors (MFxx, malfunction) would need to be sufficiently controlled according to ASIL.

The architecture metrics (single-point fault metrics (SPFM) and latent failure metric (LFM)) would result from the safety architecture and would be a mathematical function of the failure rates (MFxx) and the implemented safety mechanisms (DCxx).

The top failure metric (PMHF) has to be budgeted and distributed in different ways. In this context a budget of 1 Fit per sensor is often given per safety requirement at the sensor interface. The 1 Fit results from 10 % of the overall proportion of an item in case of ASIL D, which is budgeted for the sensor. Often, it is also indicated that there cannot be any single faults, which are more than 1 % of the target values for the overall target value of the item. This would be 1 % for an ASIL D safety goal of 10 Fit, thus 0,1 Fit. The source of these target values results from the second alternative metrics in Chap. 9 of part 5 of ISO 26262. This can lead to low target valued for redundancies and therefore to a very conservative quantification. However, since it often occurs that within the specified application space not all comparators are set on 99 % detection or certain failure conditions can't even be covered, the fit rates are still conservative even for such applications. The quantitative analysis of ASIL-decomposition can only be performed by a system integrator, since the effectiveness of the safety mechanisms and the error propagation to the safety goal can only be analyzed and made transparent by a top view from a higher architectural level, which at least consists of the redundant signal chains and the comparator.

4.4.2.9 Analysis of Dependent Failures (ADF)

ISO 26262 defines common cause, common mode and cascading as dependent failure. Dependent failure are defined as follow:

ISO 26262, Part 1, Clause 1.22:

> ***1.22 dependent failures***
> *failures (1.39) whose probability of simultaneous or successive occurrence cannot be expressed as the simple product of the unconditional probabilities of each of them*
>
> *NOTE 1 Dependent failures A and B can be characterized when*
>
> $$PAB \neq PA * PB$$
>
> *where:*
> *PAB is the probability of the simultaneous occurrence of failure A and failure B;*
> *PA is the probability of the occurrence of failure A;*
> *PB is the probability of the occurrence of failure B*
>
> *NOTE 2 Dependent failures include common cause failures (1.14) and cascading failures (1.13).*

This definition of dependability is also called the Kolmogorov's zero-one law. It is one of the laws of large numbers, since according to the definition only two cases exist, either there are dependencies or there aren't. Since we already learned that a complete independency could rarely be achieved, ISO 26262 speaks of a sufficient independency. Failures of common causes or failure dependencies between functions, which can affect through different mechanisms, are often no longer analyzable with the classical methods. In this case we can often only rely on experience. For functional dependencies we can systematically analyze a lot of things from the functional chains and their derivation in the different horizontal abstraction levels. A barrier, independent if it is a functional or technical barrier, or whatever technology it is, could be only assessed for its sufficiency or effectiveness, in the specific context and for possible failure effects (Fig. 5.59).

There are 2 definition or 2 types defined in ISO 26262.

ISO 26262, Part 1, Clause 1.13:

> ***1.13 cascading failure***
> *failure (1.39) of an element (1.32) of an item (1.69) causing another element or elements of the same item to fail*
>
> *NOTE Cascading failures are dependent failures (1.22) that are not common cause failures (1.14) (see Fig. 2, Failure A).*

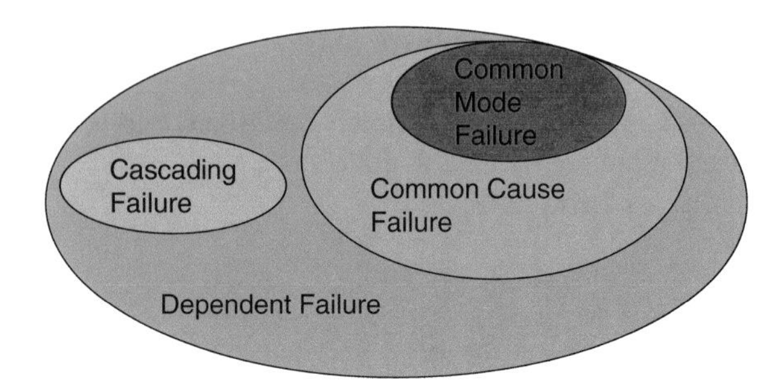

Fig. 4.59 Classes of dependent failure

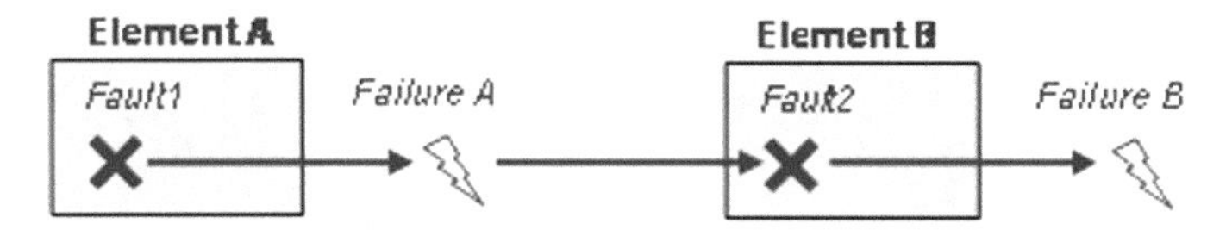

Fig. 4.60 Illustration of a failure cascade (*Source* ISO 26262, part 1)

The cascading failure is a failure, which consequently causes further failures. A cascading failure is no failure of common cause. If one of the two cascading failure is a single fault, also the other dependent failure would be a single fault depending on the operating direction (Fig. 5.60).

ISO 26262, Part 1, Clause 1.14:

> *1.14 common cause failure (CCF) failure (1.39) of two or more elements (1.32) of an item (1.69) resulting from a single specific event or root cause*
>
> *NOTE Common cause failures are dependent failures (1.22) that are not cascading failures (1.13) (see Fig. 3).*

A common cause failure (CCF) causes a failure in two or more elements that can be traced back to a cause or to a single event. A special form is the common mode failure (CMF). This failure is often traced back to the same elements, which cause the same failure behavior for a single event in both redundancy paths. This could also be the case of two different elements that for example drift in the same failure direction in case of overheating. Therefore, the redundancy would be neither reactionless nor sufficiently independent for i.e. decomposition (Fig. 5.61).

According to ISO 26262, part 9, Chap. 7, the target of the analysis of dependent failure (ADF) is to identify individual events or causes, which could lead to failure, override safety mechanism or undesired safety relevant behavior-. Following the requirements for analysis of dependent failure described in ISO 26262 would

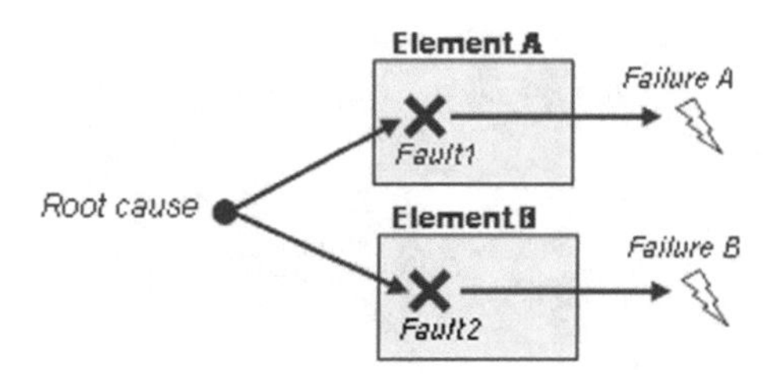

Fig. 4.61 Illustration of a common cause failure (*Source* ISO 26262, part 1)

identify many safety-critical cascades, or systematic failure which could lead to a violation of safety goals. For example an electrolytic capacitor, which is integrated in the gate path of a transistor for EMC reasons, can at age-related loss of capacity negatively influence this transistor so that it will be controlled with overly high voltage transients or a current, which is too low. As a result, the transistor can contrary to its typical failure behavior, lead to a short circuit in the drain source path. The short circuit could cause a malfunction, which then could violate safety goals. In this case even the loss of capacity in the capacitor would be a single fault, but there wouldn't be any defined requirement that the two elements need to be independent or free from interferences.. Simulations based on PSPICE (see Chap. 6 of this book) can support failure analysis regarding failure cascades as well as failure reactions for any electronic functions. PSPICE or similar simulation tools could simulate all electronic components with their characteristics also under various conditions. This means that all design related, dependent failure need to be investigated on such cascades. Often, the detection of cascades is only possible through many years of sometimes even painful experience.

ISO 26262 addresses the following objectives for the analysis of dependent failure:

ISO 26262, Part 9, Clause 7.1.1:

> *7.1.1 The analysis of dependent failures aims to identify the single events or single causes that could bypass or invalidate a required independence or freedom from interference between given elements and violate a safety requirement or a safety goal.*

The norm recommends investigating the following architecture structures:

ISO 26262, Part 9, Clause 7.1.2:

> *7.1.2 The analysis of dependent failures considers architectural features such as:*
>
> – *similar and dissimilar redundant elements;*

- *different functions implemented with identical software or hardware elements;*
- *functions and their respective safety mechanisms;*
- *partitions of functions or software elements;*
- *physical distance between hardware elements, with or without barrier;*
- *common external resources.*

According to the definitions in part 1 of ISO 26262, sufficient independency can be achieved through the absence of cascading failure and of common cause failure. For freedom of interference only the absence of cascading failure needs to be shown. This is an interesting indication in the norm, but it contradicts with the following requirement:

ISO26262, Part 4, Clause 7.4.2.4

7.4.2.4 Internal and external interfaces of safety-related elements shall be defined, in order to avoid other elements having adverse safety-related effects on the safety-related elements.

This addresses elements in general and does not somehow restrict as in the list directly related to the analysis of dependent failure. It could be that it asks for the definition of internal and external interfaces of safety relevant elements in order to avoid adverse safety relevant effects on other safety relevant elements. However, without an analysis, this requirement cannot be met. This requirement can be found in part 4, which addresses the system development. However, there is no limitation for which elements this requirement should be applied. Positively seen, this requirement refers to previous example with the capacitor and transistor, since electronic components are also elements according to ISO 26262. On the other hand, this would mean that all electronic components, even the smallest software units, would need to be checked for troublesome, harming influences of other elements. The intended function and their safety mechanism need dependencies in case of failure of the intended function, but if the safety mechanism negatively affects the intended function, the safety mechanism weakens the system. But this is again a matter of design and realization, therefore a general question, why is the analysis of dependent failure only required for ASIL C and ASIL D functions or elements?

ISO 26262 defines the following requirements, to provide some indication, how to identify dependent failure:

ISO 26262, Part 9, Clause 7.44:

7.4.4 This evaluation shall consider the following topics as applicable

a) random hardware failures;

EXAMPLE 1 Failures of common blocks such as clock, test logic and internal voltage regulators in large scale integrated circuits (microcontrollers, ASICs, etc.*).*

b) development faults;

EXAMPLE 2 Requirement faults, design faults, implementation faults, faults resulting from the use of new technologies and faults introduced when making modifications.

c) manufacturing faults;

EXAMPLE 3 Faults related to processes, procedures and training; faults in control plans and in monitoring special characteristics; faults related to software flashing and end-of-line programming.

d) installation faults;

EXAMPLE 4 Faults related to wiring harness routing; faults related to the inter-changeability of parts; failures of adjacent items or elements.

e) repair faults;

EXAMPLE 5 Faults related to processes, procedures and training; faults related to trouble shooting; faults related to the inter-changeability of parts and faults due to backward incompatibility.

f) environmental factors;

EXAMPLE 6 Temperature, vibration, pressure, humidity/condensation, pollution, corrosion, contamination, EMC.

g) failures of common external resources; and

EXAMPLE 7 Power supply, input data, inter-system data bus and communication.

h) stress due to specific situations.

EXAMPLE 8 Wear, ageing.

NOTE 1 The evaluation of the potential dependent failures plausibility can be supported by appropriate checklists, e.g. checklists based on field experience. The checklists provide the analysts with representative examples of root causes and coupling factors such as: same design, same process, same component, same interface, proximity. IEC61508 provides information that can be used as a basis to establish such check-lists.

NOTE 2 This evaluation can also be supported by the adherence to process guidelines which are intended to prevent the introduction of root causes and coupling factors that could lead to dependent failures.

ISO 26262 recommends in this context the query based on check lists, since experience widely only indicates to such failures and their effects. ISO 26262 also refers to IEC 61508, but these lists cannot be fully considered since those lists could also not be seen as complete. A more severe issue is that such list based on experiences. In case of different environmental conditions the different impacts could not be evaluated and there is no requirement, what are the assumptions for their validity.

Also ISO 26262 defines some concerns about the ability to void dependent failure:

ISO 26262, Part 9, Clause 7.4.7:

7.4.7 Measures for the resolution of plausible dependent failures shall include the measures for preventing their root causes, or for controlling their effects, or for reducing the coupling factors.

EXAMPLE Diversity is a measure that can be used to prevent, reduce or detect common cause failures.

In this case an in depth knowledge of functional dependencies and safety mechanisms of the elements is necessary for the realization. Furthermore, ISO 26262 also indicates that a fault tree analysis or a FMEA could also provide information on dependent failures. If the inner structure is unknown, dependencies can be proven with the Kolmogorov-Smirnov test. With the help of random samples it can be tested, if two random variables have the same distribution or if a random variable follows a previously adopted probability distribution. The random numbers could be systematic failure simulations. This plays an important role in medical technology or biology, but when we see discrete circuits, it will be essentially faster to find indications through the realization on why dependencies of failures occur. If systems have already multiple dependent functions or in case of using complex semiconductors or even microcontrollers such tests will be also useful, but the number of necessary test cases could lead to exorbitantly combinations. In this case, certain parameter such as overvoltage, EMC etc. can be injected and according to the reaction statements turned into possible dependencies. According to ISO 26262 the Beta-factor should not be quantified, unless the failure dependencies are based on random hardware failures, which are single- or multipoint-faults. The mayor effect on dependency and specially the criterion for "sufficiency" based on systematic faults and their average of random hardware faults are quite small.

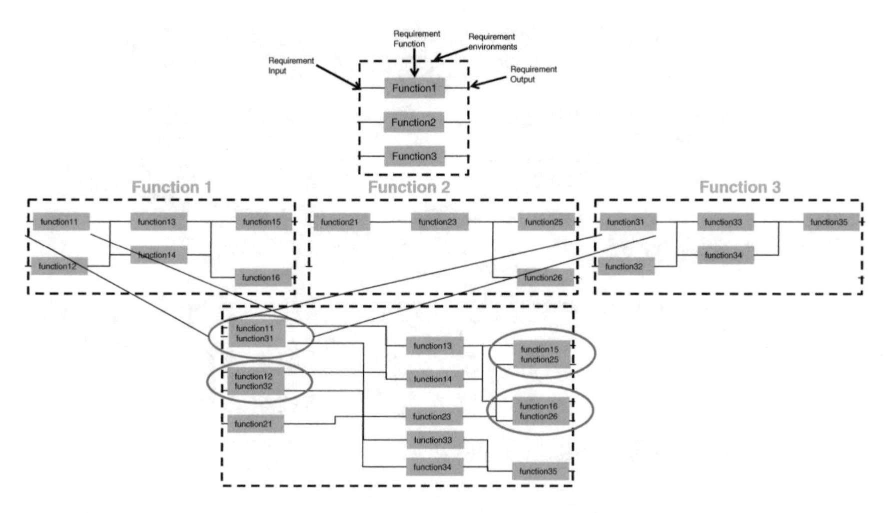

Fig. 4.62 Dependency in the lower abstraction levels by allocating derived functions on common elements

Furthermore, the dependencies are often relying on operating temperature and other environmental noise factors, so that just a factor becomes easily a huge array of data.

Identification of common resources and especially dependencies due to noise factors such as temperature or other stresses are only determine by tests based on the final realized product. The following figure (see Fig. 4. 62) shows that if in the lower levels, in particular for the realization, common functions, resources, energy sources or physically close realization elements are used, no indications will be found to that in the higher levels of the architecture. The needed information, do not derive from the functions in higher level, so that deductive analysis approaches would fail. This also applies for the software, which is processed in the same task, by the same logical processing unit or core, as well as for two electronic components, which could lead to the same failure behavior due to noises such as heat. For example if the reference signal increases in the same way like the measured signal due to heat, the comparison remains "true". Therefore, such simple comparisons will not be a useful measure in this case. For most realizations it can be excluded that information on two different electronic components is incorrectly changed at the same time, so that both they provide the same incorrect value. Consequently the probability of error propagation does not based on random hardware failure, it based more on the fact that the common case effects in the same direction at the same time-interval. By reducing the time-interval, the probability could be lowered (Figs. 4.63 and 4.64).

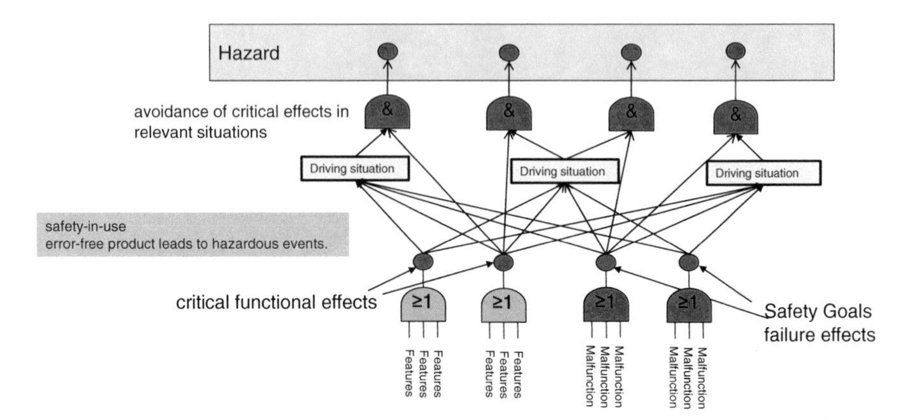

Fig. 4.63 Event tree for positive and negative effects

Single 2oo4d architecture similar to architecture from IEC 61508, Part 6 (Ed.2011)

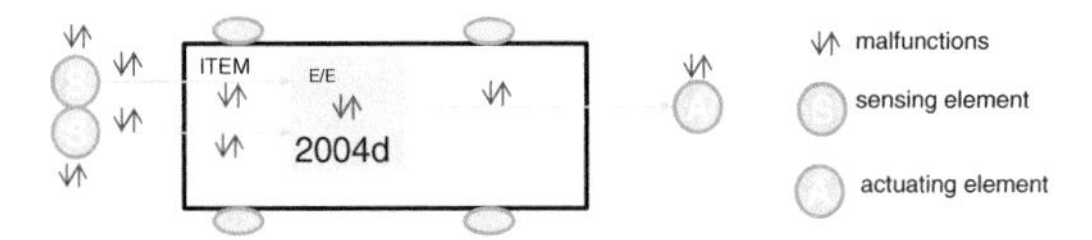

Double independent implemented 1oo2d architecture similar to architecture from IEC 61508, Part 6 (Ed.2011)

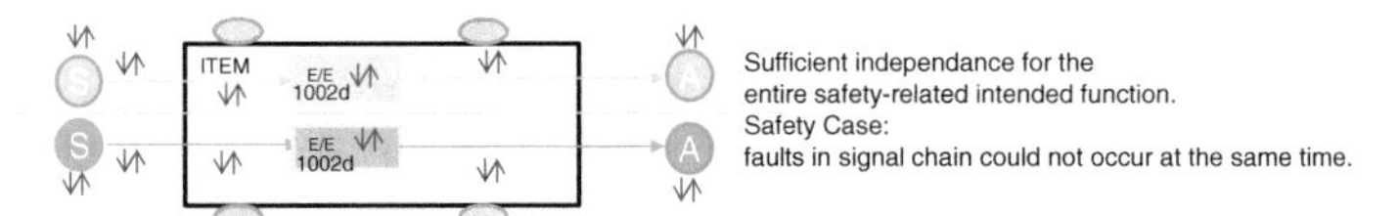

In airplane industry such an architecture called also Dual-Dual Architecture

Fig. 4.64 Independent redundant architecture for safety-related intended function

4.4.2.10 Safety Analysis in the Safety Lifecycle

In the development of the functional concept according to ISO 26262 at the definition of the vehicle system (ITEM) the first analyses (see Fig. 4.65 analysis phases) are already required, since they help to describe a risk-free intended function. But the Item definition is the only work-product without verification.

This means that we are looking for a method for the analysis of functional and operational safety. The normal operation condition for the standard road vehicles and also the basic functions are well established. Also the road traffic regulations and the considered coexistence of people with vehicles are world-wide established. They are in small details different, but mayor cornerstones are harmonized such as the "**Vienna Convention**" which defines for example, that the driver is responsible to control his vehicle. Due to today's discussion about "**automated driving**" or

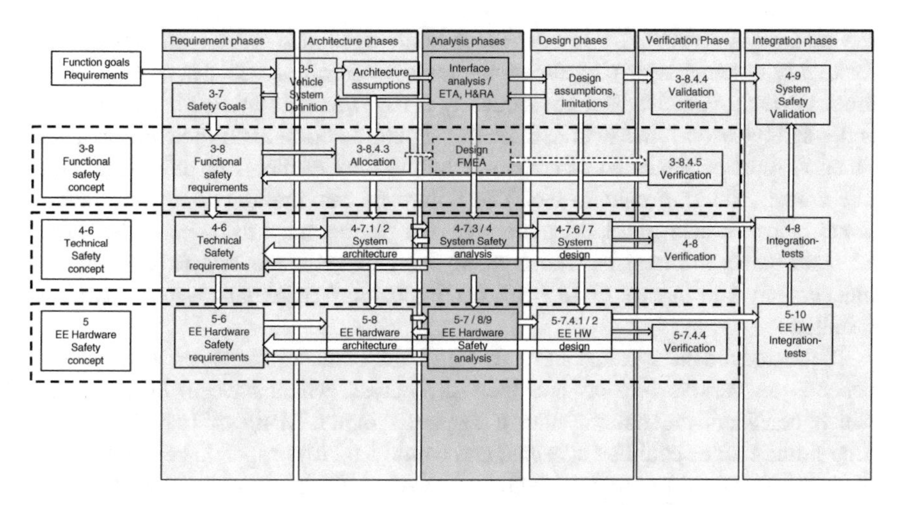

Fig. 4.65 Phases of safety analyses in the safety-lifecycle for product development on system and hardware level

even "**autonomous driving**" also other main emphases have to be considered. ISO 26262 had excluded functional performance but how could a safety case argue to safely brake a driving car within the defined limits? The following topics have to be considered:

- **Functional inadequacy**
- **Safety-in-use**

Consequently fail-operational systems need to be considered.

Safety-in-use considers that the intended function since it operates or behaves correct doesn't lead to any harm. The classical failure analyses cannot be considered for this analysis. Therefore, we rely on the positive analyses. In this case, particularly the behaviors of the intended functions, within its typical environment have to be analyzed as a positive approach. Generally, in this context we would see the classical event tree analysis (ETA). Based on deductively determined malfunctions and, in opposite to the general Hazard&Risk Analysis according to ISO 26262, effects of intended functions, within relevant critical driving situations.

As a consequence we need a detailed analysis also of the intended functions similar to the malfunctions as part of the Hazard Analysis and Risk Assessment. Different to the malfunctions which are assessed by the parameter S, E and C, the critical characteristics of the intended function need to be iteratively modified unless they could be considered as sufficient risk-free or safe. In case of a verification of the Item Definition, the safe intended function could be analyzed and confirmed. If the intended function itself is safety related like "Steering" and "braking", legal requirements like ECE R13 (or FMVSS 135) or R79 (or FMVSS 203, 204) give binding requirements for their homologation. Especially in ECE R13 requirements for the entire brake system (it is the ITEM) and its degree of redundancy is defined. Up-to-now those homologation standards are not released for highly automated driving functions, remote controlled vehicles or even autonomous driving vehicles. But consequently if the intended function is safety related and a safety case in line with ISO 26262 should be considered the entire intended function must be analyzed of course including they elements of other technology like wheels, steering column etc. The failure modes, malfunctions and malfunctional behavior also of all actuators and also sensors have to be controlled by an adequate safety mechanism. Consequently a completely independent (or even more independent than in case of an ASIL decomposition) redundant system have to be installed.

If the redundant systems have to fulfill the same performance requirements, depends on various impacts and the risk analysis which have to be performed, which considers the risk in such a degraded mode. Many of today´s accepted limp-home-modes could be accepted and would lead to acceptable behaviour of the degraded function. In order to define and analyze the boundary of the ITEM also the boundary analysis as described in the Ford FMEA handbook could be considered. On the Item or vehicle level the intended function could be defined as an "Ideal Function" in the context of P-Diagrams.

This event tree analysis or the improved Hazard Analysis needs to be continuously checked throughout the entire product development, since all new malfunctions, all changes of the environmental conditions as well as changes of the functions and the design, can lead to new effects in certain driving situations (see Chap. 4.2.1, Preliminary Hazard (Risk) Analysis, PRA). Even if these forms of the event tree analysis allow considering hazards from the correct behavior of the vehicle system, it will not be a sufficient method to show the safety-in-use of a system.

A benefit of this systematical analysis is that it shows a direct transition to the system safety analyses (e.g. RBD, FTA, FMEA). For the verification of the functional safety concept (requirement of ISO 26262, part 3-8.4.5) a more in depth fault tree analysis, reliability block diagram or a draft System-FMEA based on malfunction should follow. In the draft System-FMEA all potential malfunctions of the logical elements, the effects of the considered logical elements on each other and the deviating environmental influences (from a boundary analysis) could be investigated regarding their potential to violate safety goals. Regarding malfunctions in this case we could argue with a completeness of the safety goal coverage based on top-failure of Fault-Trees or the top-failure from FMEAs. Correct and sufficient specification of the functional safety requirements and their verification can occur through a deductive positive analysis. The analysis can be applied all the way to the lowest level (software or hardware design) and the corresponding malfunctions are complemented in the respective levels. The safety analyses should be planned in a way that they are comprehensible and consistent systematically through all levels from the safety goals to the lowest relevant failure causes in the hardware or software realization (compare Fig. 4.65). If all questions are sufficiently and completely answered, the analysis of the technical safety concept can follow. The deductive analysis (requirement of ISO 26262, part 4-7.4.3 for ASIL C and D recommended for ASIL B) thus supports the verification of the technical safety requirements (requirement of ISO 26262, part 4-6.4.6) regarding systematical failure and their allocation on technical elements, which build the basis for the system design. The following aspects can be evaluated:

- Are the technical safety mechanisms completely derived from the functional safety mechanisms?
- Have all possible malfunctions of the technical elements and malfunctions from the effect of the technical elements among each other been considered?
- Have all functional dependencies regarding functions, malfunctions and failure behavior as well as technical dependencies (for example common resources, energy, technical elements, which need to support multiple functions) been considered?
- Are all safety mechanisms completely described by technical elements? (Complete description of safety mechanisms regarding a horizontal abstraction level including all technical interfaces)
- Are the technical elements regarding inputs, outputs, the relationship between input and output, environmental conditions, permissible environmental conditions, variants and configurations completely described?
- Is the error propagation through the failure simulations (failure injections) comprehensible?
- Are the validation criteria described suitable to show the fulfillment of the safety goals?

If all questions are sufficiently and completely answered, the system design can be seen as completely and sufficiently specified. As a result, the system would need to be capable of sufficiently implementing the necessary safety mechanisms. The technical safety analysis (requirement of ISO 26262, part 4-7.4.3) should be considered as inductive analysis based on the characteristics of the technical elements in their integration environment (vehicle environment). As a recommendation, all requirements for the logical, functional and technical elements would need to be analyzed deductively. For this case the positive analysis would be completely sufficient. The aim of the deductive analysis is to identify the necessary characteristics but not to verify their values (or parameters). The "special characteristics" are also determined in the FMEA methods (because also these methods are not seen as sheer inductive analysis by FMEA) according to VDA or AIAG. ISO 26262 also addresses the "safety related special characteristics" for production related safety activities or safety requirements of the mechanical (or other technologies) elements. To develop a plug-connector for an ASIL B signal according to an ASIL, will not be sufficient information for a developer. The characteristic would need to be clearly stated as safety characteristics and sufficiently failure tolerant, reliably and robustly designed. Generally, in the automotive FMEA standards the "special characteristics" or other product or process characteristics are identified in the lower level of the Design-FMEA. Process characteristics are characteristics that need to be safeguarded though the production processed. Product characteristics are characteristics that are secured through constructions but need to be checked in the production. These product or process characteristics are handed over to the Process-FMEA (and thus to the production control plan). As a result, it can be made sure that the required characteristic is covered against constructive failures and production failures or sufficiently robustly designed.

For the deductive analysis only potential failures should be considered that are identified through those (important) characteristics, which they can negatively influence for the realization or implementation of safety relevant functions. This also applies for the functional limitations, which result from the functional design and design limitations through a higher abstraction level or the integration environment. Constraints which also have to be broken down from the Item Definition down to even the structure of semiconductors could be formulated similar to requirements. The verification of technical characteristics and their error propagation can only be performed through an inductive analysis. Verifications of constraints are more difficult to be verified, since only know negative impact could be assumed. This means that in the deductive analysis characteristics and constraints are questioned, which are necessary in order to implement the function in the investigated system (the considered logical and technical elements in their required functional behavior) as intended. These characteristics thus have the character of a requirement. For the inductive analysis we start with all technical elements and thus the characteristics, which are necessary in order for the system to carry the function, are confirmed. The other characteristics and those that result from the behavior of technical elements among each other must no influence further characteristics in a way that they are unable to discover the required function.

In the positive analysis all characteristics, which are important for the realization of the function (or the safety mechanism) its dependency to other functions and its derivation through multiple horizontal levels can be determined.

The negative perspective, simply converted according to DeMorgan's law, leads to highly complex, logical correlations. The 3 positively illustrated sub functions together build the main function. However, at the negation the failure of a partial function can lead to the failure of the main function, as well as all combinations of possible failures and failure behavior. For the planning of the degradation concepts, only the failure of the function needs to be considered, since the degradation cannot only evacuate individual failure. This is especially important, if we speak about highly available safety systems. In this context the aim is to retain the function in a reduced extent. However, the error propagation can't even be assessed without the technical realization. Vibrations and oscillations can only arise, if they are evoked by, for example, inductivities and capacities and if there is sufficient energy to significantly disrupt such a system. The same applies to drifts of signals: drifts upwards need energy. If this energy isn't available it also can't lead to a drift of signals. Energy can be derived for a signal through the realization (e.g. cross-talking), which can even lead to drifts into the negative, which maybe hasn't even been considered. ISO 26262 does not address which method for analysis or verifications has to be applied but other industries see an accompanying deductive analysis during development of requirements as the most target oriented approach.

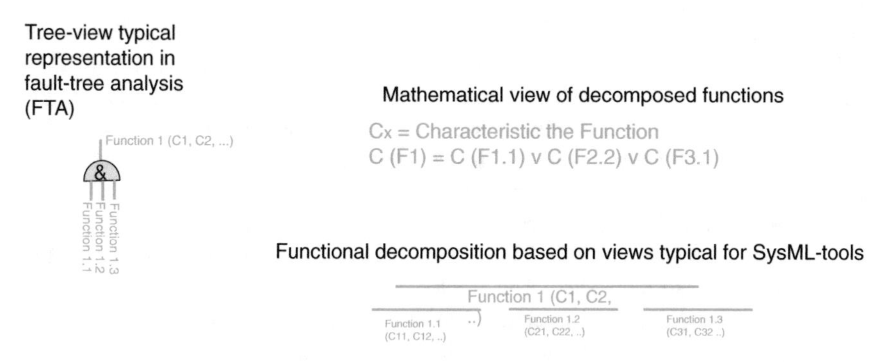

Fig. 4.66 Positive function analysis or decomposition; Illustration of a functional dependency as a tree, a line diagram or a logical/mathematical view

A hierarchical functional decomposition could be applied similar to a positive "Fault Tree Analysis", if be applying with the lowest function "DeMorgan's law", a complete set of malfunction for the lowest malfunction could be evaluated. If these malfunctions would be analyzed from the bottom to the top (potential violations of safety goals) verification for completeness could be demonstrated. This bottom-up approach could be done by means of an FMEA, so that additional safety mechanism could be defined as measures of the FMEA (Fig. 4.67).

The same correlations can be found for the development of components. In the software it is useful to analyze deductively and functionally even during the development of requirements and therefore determine the key characteristics of the elements, which are necessary for the correct implementation of the function. After software design is finished an inductive analysis should follow. It should proof whether all systematic failure, which still remain in the software, are covered by sufficient measures.

A different approach is used for the electronic development. However, also in this case, it is still useful to deductively determine the characteristics, which are necessary for the realization of the function. A lot of FMEA standards have also recommended the same. Generally, this will happen in the context of a

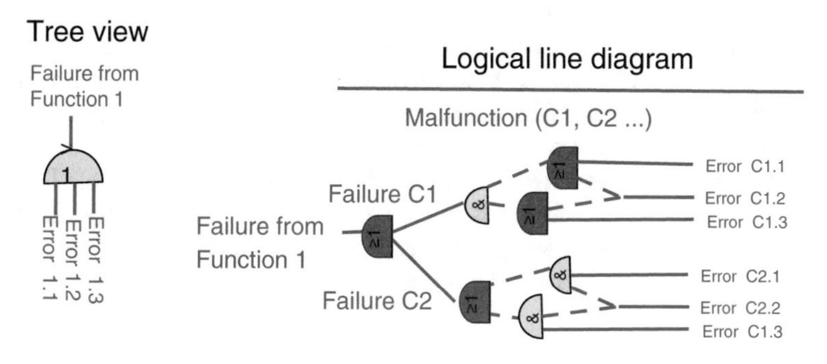

Fig. 4.67 Negation of the functional analysis of Fig. 4.66

Design-FMEA, whereas these key characteristics will be determined in the middle failure level and causes from design or in the production will be examined. If we want to remain within the classical three ‘cause of failure – type of failure – failure effect sequence’ one level is often missed. Therefore, the key characteristic is often also defined on the cause level. In this case, the Process-FMEA would start below the causative level of the Design-FMEA. The Design-FMEA mainly meets the requirements of ISO 26262, part 5, Chap. 7 but for the identification of single faults, multiple faults and safe failure a consideration as System-FMEA on component level is also necessary. The implemented safety mechanisms are then the measures in use (or during operation). The architecture metrics, required in ISO 26262, part 5, Chap. 8, then use the quantification of the failure modes and the efficiency of the safety mechanisms (diagnosis coverage against single faults or latent faults) as standard for the assessment of the safety architecture.

The cut-set analysis, required in ISO 26262, part 5, Chap. 9 for the development of PMHF (probability of the harm of the safety aim through random hardware failure) can only be performed based on the realized design. At this point information is received of the architecture metrics and the analysis of dependent failure (ADF). Already for double failure we can question whether they are even able to cause a quantifiable violation of safety goals. If it is about an independent random hardware failure, the influence to safety goals is calculated by the multiplication of both failure rates (Fit values), which results in very small values. If both failures are not independent (which is very often the case for automobile electronics), the degree of dependency (see Kolmogorov’s axiom) determines the probability that those two failures occur together and have the potential to violate safety goals. This would comply with the quantification of the beta factor. If the dependency refers to technical electronic elements, ISO 26262 requires designing it in a way that the degree of coupling-effects is sufficiently low. The failure proportion of this electronic element would be included as a single-point fault in the metrics. However, since there are often design related and thus systematic dependencies, which are often causally based on very complex combinations (cut-sets) of influence factors, ISO 26262 doesn’t require the quantification of the Beta factor but only an identification of the influence factors and respective measures in order to control these dependencies. The failure rates for the architecture metrics are allocated to different requirements in ISO 26262 than the failure rates for PMHF. For the architecture metrics the focus lies on the balance of data among each other so that we cannot justify the coverage of essential functional elements (or functions) with malfunctions, faults, errors, failure, safe failures or diagnosed failures. In order to quantify faults modes from the failure rate of electrical elements we simply use failure rates from data manuals and assume a generic failure distribution (for example Birolini). For PMHF the focus lies on the failure propagation and the necessary influence factors. In this case it is also required to determine the failure rates, which are based on the realization as well assess the influence at the cut-sets.

4.4.3 *Safety and Security Error Propagation*

Safety and Reliability follows similar principles for the failure analysis. Especially dependent failure and their analysis did show that the typical sequences of fault, error, failure or failure cause, failure mode and failure effect are not always applicable. Similar challenges are affecting security analysis. Measures to control the different security threads like Integrity, Confidentiality and Availability show different relations to their possible effects and effectiveness.

Deep Medhi provided a key-note where he presents a common "Dependability and security model"

The possible threats (mainly security) had been defined on a level of faults, so that further propagations lead to errors, failure and accidents. Unauthorized access to data and also unavailability had to be considered also as an accident in this case.

Safety is defined as "Attributes" as well as also availability, confidentiality, integrity, performance, reliability, survivability, and maintenance. Similar considerations are in railway standards with "RAMS"-Approach (Reliability, Availability, Maintainability, Safety) and shows also similar ideas like (Fig. 4.28. Basic principle of FMEA) the chapter about error propagation principles. Possible measures, called means, are also similar to possible measures in an FMEA (see also Fig. 4.41):

- Fault/Intrusion prevention
- Fault/Intrusion tolerance
- Fault/Intrusion removal
- Fault/Intrusion forcasting

This shows that similar analysis approaches for safety and security could be applied, but that there no one2one-relation to be expected.

Security analyses are as similar to safety analysis as reliability analysis. By considering the such well-tried analysis principles, all kind of threats, critical impacts or any other unwanted event to systems or products could be examined.

4.5 Verification During Development

During the development of the product ISO 26262 asks frequently for verifications. Most likely always, if a development activity relies on the input of a former development step. In the descending branch of the V-model verifications are always required at interfaces of horizontal abstractions.

In this context, the verification is seen as the completion of a higher level activity and the lower level activities usually begins with a requirement analysis. ISO 26262 considers also tests especially in the lower horizontal abstraction levels, particularly during component design as verifications. However, methodical, the method for the correct derivation from a higher level would need to be a validation. What is important though is that this verification regarding correctness, completeness and consistency is based on the same consistent verification method. Most likely allocations are done before the verification, which means that requirements are allocated to the underlying levels elements. In this case the relationship between those two levels needs to be analyzed. The verification does not initiate process iteration only if the results of verifications are flawlessly positive. Depending on how deviations are assessed during a verification and what measures are initiated for the repetition of the verification, we need to go back to the corresponding previously activity. These could be the requirements, architecture, design or also the test case specifications within a horizontal level as well as a jump back into another horizontal level (for example from a component into a system, or even on vehicle level, so that safety goals could be affected by changes). The first activity during verification of requirements should be a requirement analysis. Therefore the question is: Are the requirements of the lower level derived from the requirements of the upper level or from constraints, architecture or design of the higher level? As already described in the introduction of Chap. 4, ISO 26262, part 10 (Figs. 7 and 8, see also Chap. 4, Figs. 4.1 and 4.2) renounced the maturity level description (system design V1.0 etc.). The figure originally wanted to portray the information that through the different levels, the design always includes more and more detailed information and especially the relevant design characteristics become more and more plausible during any iteration. However, those should be tested or verified before they are passed on to a further user of this information. This shows that there are different ways to develop work results. Basically, we distinguish between a requirement specification and a design specification. However, there are different manifestations and definitions as to how both specification types can be structured. At this point, the requirement specifications should provide the general conditions, which are necessary as foundation for the design. The design specification describes the implemented characteristics, which can be measured by the product. A requirement specification defines "How it should be" and the design specification defines the "how it is designed". We now also reach the performance limits of a process models.

Does the verification really only happen during the development of the requirements? Is the development of requirements completed before the realization? Obviously not! When the result is validated and all requirements are correctly implemented at the product, there are always new aspects that can occur in the usage phase, which haven't been sufficiently considered.

Also in this case, constant iteration loops occur and because of today's short innovation cycle, products are often only mature after years of their usage and each change also becomes a risk for other characteristics. This of course is not acceptable when it comes to the safety characteristics of a product. It is true that an inexperienced development team often doesn't know the influence factors, but an experienced team can also make incorrect assumption. Unfortunately, there are certain amounts of risk even in the approach itself. If requirements are systematically developed and properly derived according processes, the known influence factors will also be incorporated. If experienced people perform these analyses, some aspects will also be included in the analysis, which go beyond the requirements and the experience of the designer. At the verification certain levels of experience can also be incorporated through the test planer. Also through the individual analysis or verification methods systematically complementary influence factors are considered. However, it is hard to say or maybe even impossible to assume that all influence factors will be considered or even all application scenarios and relevant conditions. If we now have a test case for each requirement (according to the process model derived e.g. from SPICE), which shows that the requirement is implemented correctly, there will certainly be doubts regarding the significance of the tests. How many tests are necessary, will depend on a variety of factors, and even on the diligence, with which the requirement analysis has been performed at the beginning of the processing of the abstraction level. Requirement and design specifications should formally be stored in a way that there is actually only one parameter in the requirement but through the further derivation of the design information essentially more parameters are developed. Otherwise, the lower levels cannot be sufficiently supplied with information. The most concise example occurs at the hardware software interface. The design of the microcontroller provides the essential software requirements, not the requirements from the upper levels. Therefore, ISO 26262 requires the verification of all SW requirements but it is not possible to directly derive them from the system requirements, which are allocated to the software. All basic structures of the microcontroller have to be included in the requirements of the hardware software interface (HSI), which is often unable to fulfill the relevant system requirements for the software components. However, this is a very concise example, but definitely not an unusual exception.

Besides the safety analyses and tests, more and more verifications are necessary for the determinations of the safety maturity for the product under development. At each organizational interfaces and all horizontal interfaces as well as the in between elements, all characteristics should be verified at the end of the development. In general verification could show the fulfillment of requirements, from a methodology point of view you only get answer if the targets or goals are fulfilled could be only shown by validation. The activity to validate the correctness of requirements, by evaluating higher level requirements or constraints to their correct derivation to lower level requirements called ISO 26262 “verification”.

4.6 Product Development at System Level

From a Marketing point of view, products are means that can satisfy a need of a customer and thus generate a benefit. This benefit can be materialistic or unmaterialistic, but also functional or non-functional. The core of the product will for example be a technical benefit and additional benefits are perceived by the customers themselves (this could be for example quality characteristics—how beautiful, impressive etc.). Furthermore, the user of a technical system will also face burdens, for example, a product needs energy or dissipates heat in order to fulfill its purpose. Therefore, it is impossible to only look hierarchically from the top to the bottom. We now have to deal with the characteristics of components, with which we want to compose the system, and check, which characteristics and requirements comply with the functional concept and the further stakeholder requirements as well as the technical safety concept and which additional characteristics create a positive benefit (especially regarding the performance requirements) and which are an unintended negative burden.

Basically, there is always a certain discussion regarding the border between architecture and design. There are advantages and disadvantages for one or the other opinion, but it is more difficult for the product development if this isn't defined at all. This is why the following principles and analogies are used.

Architecture determines the structure and therefore the interfaces of the considered elements. Elements can be functional, logical or technical elements, which behavior among each other results in the desired functionality. A system is a limited amount of functional, logical or technical elements, which realize desired functions or functionalities through their interactions. A system should be also limited through the horizontal abstraction level where characteristics and also the technical behavior are specified. The characteristics and the described technical behavior can be specified in natural language as well as through semi-formal and formal notations. A model is therefore mostly a description of specifiable characteristics and behavior of elements under consideration of architecture. Design is also an illustration of technical characteristics of different perspectives. The following analogy is developed from these specifications for the horizontal level:

- System design is therefore the illustration of technical characteristics and components, from which the system is composed. The system design specification describes the characteristics, which result from the interfaces of components. The components can also be made of functional, logical or technical elements and have characteristics that result from the interfaces of these elements. These elements need to be specified as well.

This would in general be done within the component specification. Technical components consist of mechanical, electrical or software elements, whereas the combination and the selection, which elements belong to which component, represent a design decision. A component is therefore a subsystem of the system, which forms the differentiated characteristics from the interaction with other components.

- Mechanical design is therefore the illustration of technical characteristic of mechanical elements (components), of which a mechanical system is comprised. The mechanic design specifications describe characteristics, which result from the interaction of mechanical elements. The mechanical elements can consist of logical or technical elements and have characteristics, which result from the interaction of these elements. These mechanical elements need to be specified as well. This would in general be done within the component specification. Mechanical components consist of mechanical elements, whereas the combination and the selection, which elements belong to the component, represent a design decision (Fig. 4.68).

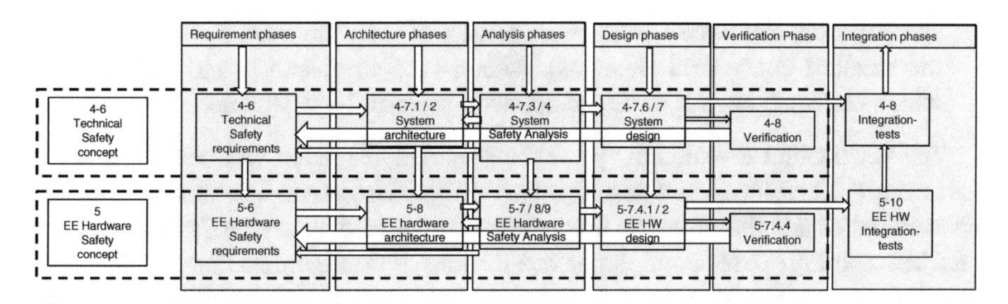

Fig. 4.68 Information flow in the system and EE hardware development derived from technical safety concept (TSC)

- Electronic design is therefore the illustration of technical characteristic of electronic elements (components, electronic parts), of which an electronic system is comprised. The electronic design specifications describe characteristics, which result from the interaction of electronic elements. The electronic elements can consist of functional, logical or technical elements and have characteristics, which result from the interaction of these elements. These mechanical elements need to be specified as well. This would in general be done within the component specification. Electronic components consist of electronic elements (parts are the smallest elements for discreet electronic, for semiconductor sub-parts are defined as logical or functional units to sufficiently describe the behavior and the relevant characteristics), whereas the combination and the selection, which elements belong to the component, represent a design decision (Fig. 4.69).
- Software design is therefore the illustration of technical characteristic of software elements (elements realized in software), of which a software system is comprised (only software based system). The software component specifications describe characteristics, which result from interfaces of software elements. The software elements can consist of functional, logical or technical elements and have characteristics, which result from the interfaces of these elements. These software elements need to be specified as well. This is often called a software design

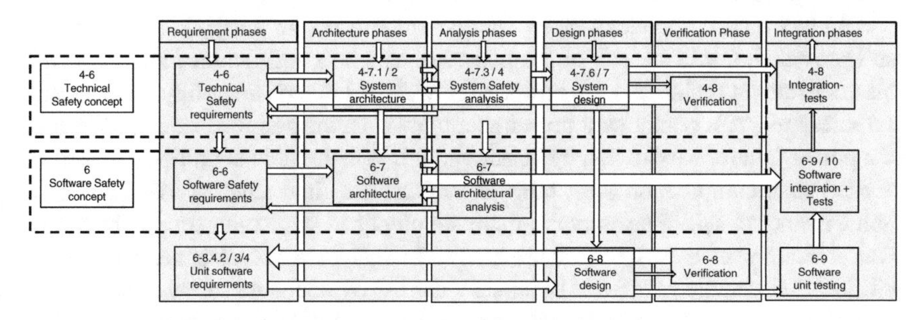

Fig. 4.69 Information flow in the system and software development derived from technical safety concept (TSC)

specification. Software components consist of software elements (SW-units are the smallest considered elements), whereas the combination and the selection, which elements belong to the component, represent a design decision.

The verification reveals how good the specified characteristics of the respective designs are. Therefore, each design decision should be verified so that the specified characteristics and therefore the implemented requirements can be described as correct. Traceability is a result of proper verifications, if the arguments for completeness, correctness and consistency are transparent, sufficient traceability would be in any case available. A general verification for correctness can only be performed through falsification, especially if constraints are contradicting performance requirements.

In the deductive safety analysis all possible variances and consequently the entire specifiable space should be analyzed. In the inductive safety analysis the specified elements are considered at the respective horizontal abstraction level and the possible error influences or impacts are evaluated. As a result a systematic falsification of the specified space could lead to completeness regarding possible error behavior. Influences and combinations, which the developer cannot imagine or not systematically evaluate, are also not verifiable. The characteristics of the product should be ensured at the end of such horizontal development activities after their verification.

Interfaces of elements are always given by the nature of electronic systems. Those interfaces must be specified in order to assure, that systems could realize the intended functions. A mere software component only exists because it is defined like that. The low level driver (e.g. MCAL; microcontroller abstraction layer), which read the information from the microcontroller hardware and provide the data interface to further software components, these software elements built the hardware-software-interface (HSI). Mere electronic parts could not be considered without their interfaces to mechanical elements, such as PCB, connectors etc. This means, no matter at what horizontal abstraction level electronic are considered, there will always be mechanical intersections. Even the bonding (connection between the silicon and the pin) in a microcontroller or ASIC primarily depends on the production process, which creates the mechanical connection. The example of the hardware software interface shows how deep we have to go into the details of the components in order to ensure sufficient coverage and sufficient unique level of specification.

All electronic components need energy (often a power supply); if this energy isn't provided, it will be one of the first reasons why the function fails. However, this does not yet define that the power supply should inherit the safety requirement. If a safety relevant component unrestrictedly reaches the assigned safe-state in case of a power failure, we can assume a safe failure. Whether this condition can also be reached in a combination with other errors and we thus have to assume double faults, is not yet said. Furthermore, there are plenty of semiconductors, which only work correctly in a certain voltage range. In this case we also need to check, whether this voltage range has to be safeguarded by adequate safety mechanisms. All those questions can only be answered by the deductive safety analysis and thus also up until which periphery element the safety attributes (i.e. ASIL) will be inherited. In combination with hydraulic there will also only be one function given at the system level, if there is a certain hydraulic inlet pressure at a valve. If this isn't the case, the hydraulic might not fulfill a possible safety relevant function after the control of the valve. The iterations cannot even be made transparent in the illustration of the information flow. The sensor will be seen as logical element in the first iteration and further partial elements (power supply, holder, housing, wiring, and also the counterpart at the control unit, which reads in the sensor information) will be successively complemented until a design decision has been made. Then it has to be tested, whether the technical element actually meets all requirements, which should be confirmed by the tests as part of the verification.

The inductive safety analysis will show whether further characteristics can lead to failures, which can influence or violate certain safety requirements or safety goals. Besides the functional behavior, interfaces will be enriched with technical information (geometry, material characteristics, temperature characteristics, stress behavior (robustness)). Consequently, a logical element will successively turn into a technical element in the course of a design phase. The design characteristics documented in the illustrations (layouts, figures, sketches etc.), parts lists and design specifications will be further confirmed with each verification and iteration until a clear element remains, which is sufficient for the application. From the process point of view the V-model will be turned upside down in the last design phases. The design decisions in the lower levels need to be verified and analyzed one more time, so that the tests can be performed based on secured and correct specifications for the integration in the upper horizontal levels. Since in the design we often specify a conservative assumption for the application, small changes will not have a massive influence on the upper design specifications. Generally, it is important to allow for sufficiently robust interfaces in an early design phase so that in the case of changes the dependency and the influence on other elements can be limited. This is also important when different variants are planned for the products. In this case, interfaces need to be planned for the variable elements, which can decouple the dependencies and influences to the other elements.

4.7 Product Development at Component Level

In order to meet the requirements of a hierarchical design we will need to pursue a system development approach also within the component. Since at the realization we also have to consider intersections between software, electronic and mechanics, and related tools this approach should also be continued just like at the system level. Generally, also in this context we will describe technical elements as logical elements until the last design decision. A microcontroller will never be fully described as technical element, since the functions of the individual transistors in silicon are more defined through a probability distribution than actually through technical functionalities, as we can imagine for the interaction of a bolt and nut. Almost all technical behaviors are described through more or less solidified models.

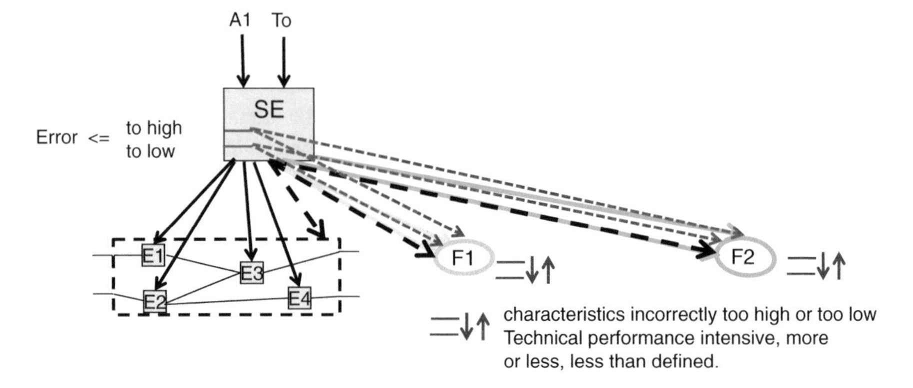

Fig. 4.70 Derivation of requirements for functions and deductive failure analysis

In mechanics, there are Newton's laws; in electronic we rely on Ohm's law until we have to use Maxwell's equation for high frequencies as a different description basis.

As a result, parallel to the derivation of the requirements for a system element to a logical element, a deductive analysis is performed, which tests how the system element changes its characteristics and behavior to its environment, if the defined characteristics are not fulfilled (compare Fig. 4.70). In the second step these logical elements are mapped on technical elements and thus design decisions are made. Those design decision and its possible faults will now be questioned and tested through the inductive analysis (from known characteristics of the design elements to possible error propagations etc.), if the influences or effects, which are determined through the analysis, are confirmed as sufficient robust (or any other quality target) or changes lead to design iterations unless no unwanted effects and the required assurance of characteristics could be accepted. Since now technical elements (which can be broken) are considered, the individual characteristics and the behavior can now be assessed through analyses, simulations, calculations and tests (which are the typical measures of a Design-FMEA).

If we now transfer logical elements to technical elements (compare Fig. 4.71), we realize that without finding a 1 to 1 allocation the number of interfaces could rise exponentially.

All technical interfaces

- between environment and technical elements
- between technical elements and functional interfaces
- between environment and logical elements
- between logical elements
- between technical elements
- between logical and technical elements

All those interfaces need to be considered, since all characteristics and their possible failures at these interfaces and all expected behaviors of the elements among each other can lead to failures, inconstancies or deviations towards the

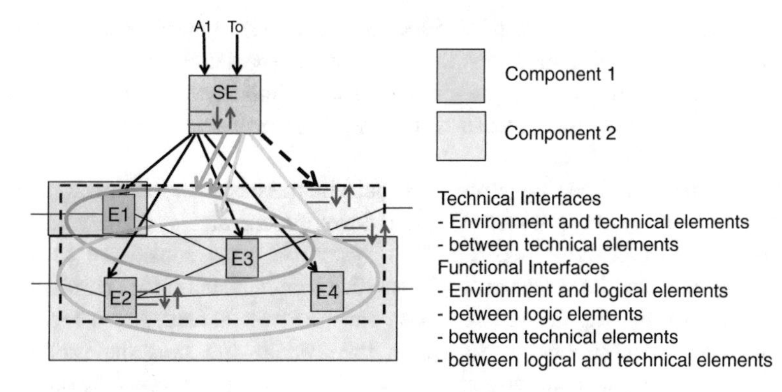

Fig. 4.71 Vertical functional decomposition and technical interfaces and their potential malfunctions

overlying requirements for the system element. The tolerance that possible failures or deviations don't lead to a harm of the overlying requirements can be considered as safety robustness.

Drawings, circuit board layouts, parts lists, data sheets and design specifications describe the expected view at the realized product. In the context of product design, characteristics of the product should be documented as part of design specifications. The structure of the architecture for the product determines the intersections for such considerations and lead to the structure of the design specification. For the product liability it is important to also indicate risks of the handling or usage of products in the product description, which is another aspect of safety than functional safety.

The automobile industry uses the Design-FMEA to analyze the design and the determining of risk of the design faults. The Design-FMEA is a risk-based approach, which analyzes the design of the components mainly to evaluate measures during development. The same approach could be also considered on system level, these in some standards are called a Design-FMEA on system level.

The metrics, such as the risk priority index, indicate whether the design is assured by sufficient measures. System-FMEAs (seen as methodology) analyze mainly the architecture and consequently primarily the interfaces. This is why the Design-FMEA often goes to deeper level of abstraction.

Ford´s FMEA handbook also requires a Design-FMEA on system level, in order to ensure that the components interfaces are designed correctly. Aircraft standards require similar approaches; also the Product-FMEA according to the VDA standard could provide a similar interpretation. Managing of failure interfaces can often be challenging, since often multiple suppliers need to be coordinated under the directions of OEM. ISO 26262 requires incorporating the coordination as safety activity in the development interface agreement (DIA).

The Design-FMEA is used to identify product and process characteristics, which need to be communicated at the interface to the plant. They are called "special characteristics" if they are legally relevant, safety relevant or economically of importance. The effective measures to control such "special characteristics" should be defined, so that they can be proven in a product liability case. In this case they also have to be documented and should be archived or recorded years longer than the estimated product live.

Such a system engineering approach should be the basis for the development of each software intensive product. Product development and system; what is the relationship between those two terms? A system is often considered to include engineering and functions. Generally, it includes elements, which are combined in a specific way so that the desired functions can be implemented. If we look at the elements and the components, of which the system and thus the product are comprised of, with the same principles in mind, we can also reach the consistency in the technical behavior and the relation to the characteristics of the components independently from the technology they are based on. In aerospace standards and information technology the technical behavior, in particular the information flows between the components or other elements, is called 'processes'. This is an important aspect for the system safety and the correct functioning of products besides the structured hierarchical classification and the design limitations. A product is more of the object of a contract, where functionality, characteristics and design properties are defined or specified.

4.7.1 Mechanical Development

ISO 26262 does not explicitly address mechanics. Most of the sheer mechanical products were clearly defined through drawings, thus specifications cannot be found for all mechanical products. The data archives in SAP are referenced to the design drawings and further documentation such as specifications are attached to these drawings. Mass production products such as caliper for brakes have a product data sheet and these are considered as sufficient complete for integration. A list of "Special Characteristics" is also referenced to these drawings. However, in correlation with electronics, we cannot completely neglect mechanics.

Plugs, housings and circuit boards are mechanical elements. It is intensively discussed where the border between the electric and mechanical element lies in a valve or engine; for the coil and windings we can see the tendency to call them electric components. For a mechatronic system we will not have a choice but to consider a system development approach, since the interfaces of a lot of different technical elements will need to be coordinated. For a mere hydraulic or pneumatic system the correct interaction or the technical behavior of the elements plays a major role. A function cannot be defined through the characteristics of the elements alone. The required characteristics of the function will only be effectively achieved, if the elements are optimized with their characteristics according to their requirement. Mechanical elements can fail, just as well as electronic elements, because of random hardware and systematic failures. However, it is not recommended to consider these random hardware failures referring to the metrics of ISO 26262. It is true that pure mechanical systems can be electronically monitored but the known databases are still too different in order to come to correct mechatronic functions or comparable failure rates.

A classical brake booster can be planned as logical element in a brake system. It is also possible to break down possible requirements to valves, springs or other logical elements according to the specified integration environment, without considering these elements as technical elements. A spring in a mechanical model can be described purely through the spring constant and statements can be made on the sufficient spring typical parameter. However, if we have to make statements concerning the use, aging behavior, stress or elasticity of the spring, we would need to consider the spring as a technical element. It could be questioned, if it is necessary to have specifications in natural language and data sheets for the entire components or partial elements, in order to ensure the safe function, such as ISO 26262 requires for electronic elements and components. The influences of a spring to a brake booster need of course geometrical consistency of the data for correct functioning, but the influence to elements of other technical elements especially to software could be only described on a functional way.

Especially for hydraulic and pneumatic functions there are standardized descriptions or specifications, which provide significantly more and more precise information than a requirement in natural language. Nevertheless, the mechanical components will be analyzed regarding systematical failure in order to question the sufficient design (for example by means of a Design-FMEA) and also to analyze the interfaces and their correct behavior in the customer's environment in the context of for example of a System-FMEA. Of course, the derivation of the requirements and design decisions can be supported with deductive analyses. Particularly the selection of suitable partial components can be supported with a deductive analysis. Generally, functional correlations will be easier to analyze and illustrate than software intense components. The details are easier to observe by experiments (e.g. DoE, Design of Experiments) or other test methodologies, and automotive industry is very experienced in the verification of mechanical components.

4.7.2 Electronic Development

It is not necessarily the traditional way to use a system development approach for the development of electronic components. In this context, the digital bus cable is seen as electronic connection and the power supply cables and their plugs etc. as electric connections. Since the discussion, whether something is electric or electronic, does not cause any changes in the requirements or benefit the safety, we generally use the term "electronics" as umbrella term. This should not be mistaken with the distinction of electric safety and functional safety, since also failures of electronic components can lead to hazards, which are associated with electric safety. This applies especially for power electronics in the voltage range of over

60 V. For higher voltage than 60 V DC or 25 V AC, legal requirements have to be considered due to touch protection. However, for the description of the requirements for electronics ISO 26262 still chose a V-model as reference model. The question for the horizontal abstraction level is now: Where does the electronic development start and where does the system development end?

The **hardware—software—interface (HSI)** requires already a very detailed level of abstraction and at this level it will be difficult to have complete signal chains for the entire ITEM functions or other functions on vehicle level. The HSI is a typical example where we could not have complete system engineering on one horizontal level of abstraction. In case of interface like the HSI, the functions and also the safety mechanism affect the system on such deep level of abstraction, the that also analysis need to be done more detailed as for example on a sensor interface with discrete electronic parts. Furthermore, the question is also if a 100 ohm resistor needs to be specified in natural language in order to ensure the necessary functional safety or if data sheets are sufficient. Resistors as measuring-shunt or as pull-up-resistors require different way of consideration during safety engineering. In this context, requirements for discrete electronic components will not be considered as safety relevant elements and a safety relevant function will only be realizable only through the correct interaction with other corresponding electronic parts or components. This is why it has to be carefully tested, if a functional requirement for electronics is not already sufficient as a safety requirement. If a RC element should work as filter or even more specific as low-pass filter with safety-related characteristics, the filter function or the required low-pass performance and the necessary time constant (T) needs to be specified as safety requirement or as safety-relevant function but not the data of the resistor or the capacitor.

In the first iteration the system requirements, design limitations, architectural assumptions and other constraints should be derived to functional or logical elements of the electronic. Architecture assumptions for the electronic, which are in line with the system limits (see Fig. 4.72) are the basis for this break-down. Base on new insights of other impacts specification of the elements gets more and more mature in further iterations. During those iterations and their verifications, the functional requirements and also the requirements for mechanical devices such as plugs, housing, circuit board, and fuses etc. have to become sufficient consistent.

The circuit board concept often based on previously developed products, internal power supply concepts are most likely a carry-over from other designs. Furthermore, we will need to think about the conductor path management for higher voltages and currents. Since housings need to be chosen very early, analyses such as thermal balance, power load (permissible short-circuit currents etc.), space requirements (housing volume, size and distance for example between pins, conductor paths, mechanical support (plug, printed circuit boards)) or energy balance need to be performed. The resulting specifications then cause most of the design limitations, which need to be considered for the design of the electronic components.

For the electronic design we need to define electronic components (see Fig. 4.73), which should be able to be broken down within the limits of the

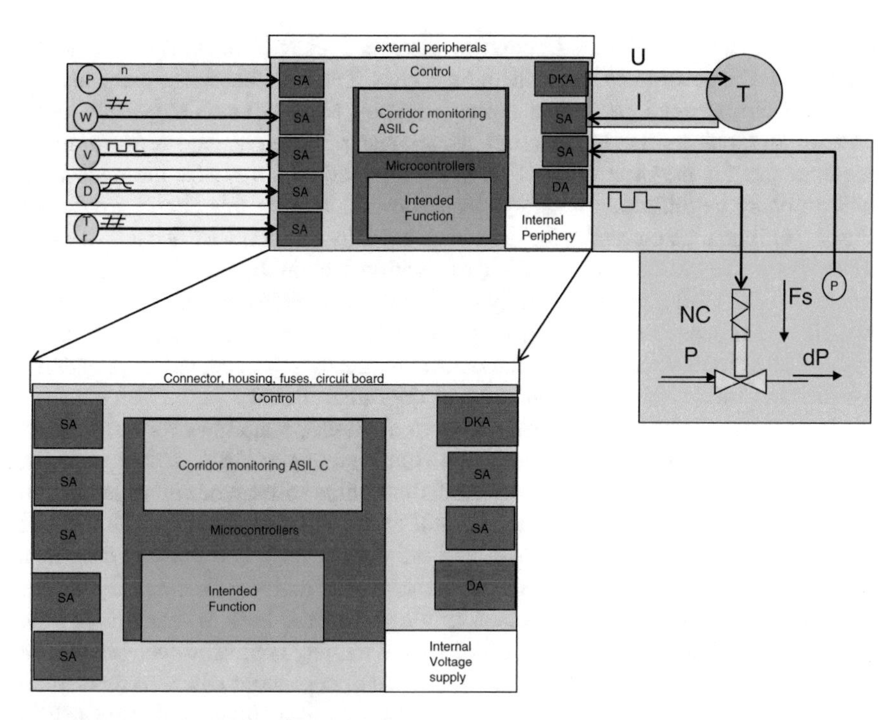

Fig. 4.72 Element based break-down of system elements and allocated requirements to electronic hardware

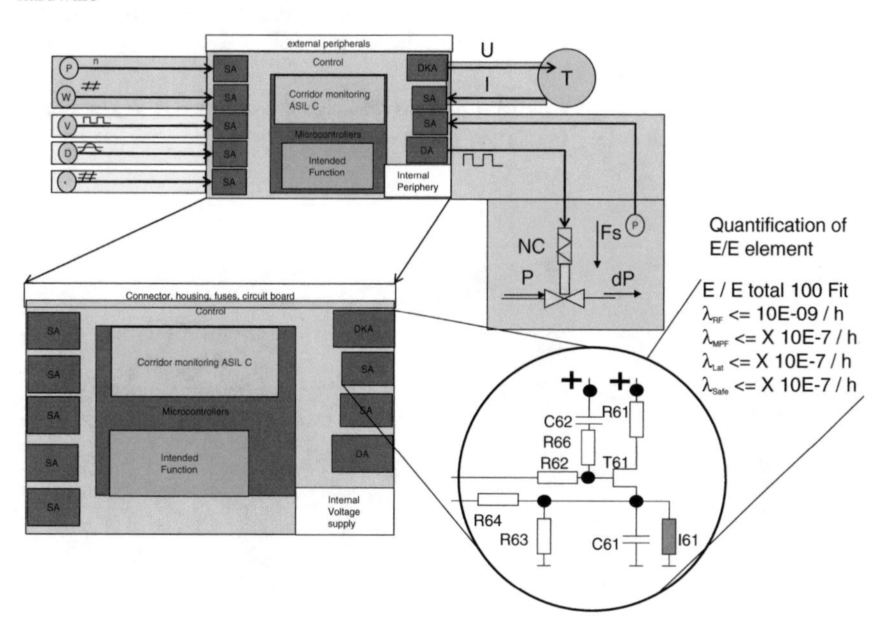

Fig. 4.73 Element based break-down of system elements and allocated requirements to electronic hardware down to realization on E/E part level

functional elements, so that electronics parts and their characteristics could be identified. The recommended realization shows that two functional elements are allocated at one circuit group. The motor winding (I61) is not seen physically in the control unit but in the engine, but as an electric component needs to fulfill requirements for the electronics. This is a simplified circuit group; the realization will probably be different for safety electronics. As long as this circuit group (see Fig. 4.74) functions correctly, the read back capacity through R64 needs to be able to measure the current in the coil (I61) within certain limits. However, if the capacitor (C61) fails, it is not directly possible to differentiate whether it is actually a fault in the capacitor or a fault of the coil. The example shows though that a separation of logical elements for the realization needs to be planned or considered for the later safety analyses or analyses of dependent failure.

As an architectural decision there are two different possibilities for the planned realization idea. We could consider these circuit groups as ASIL decompositions, since errors, which lead to safety relevant failure of the coil can be controlled in two different ways. We could control the current to the coil directly and reach the safe state by switching current off or we read the current back and react in case of a deviation between send and received current. Alternatively, we could see the control at the functional path and consider the power read back as measure. In both cases, a single fault will not lead to a violation of a safety goal. However, we will be able to identify the capacitor C61 in this analysis of dependent failure as a potential single-point fault (SPF). The coil could also be seen as a single-point fault (SPF). Since a redundant implementation of C61 cannot be recognized, the coil will not be part of an ASIL decomposition. However, it will be possible to infer from the current through the transistor to the current in the coil. For this solution two requirements are considered.

x.1 For a positive voltage control of the driver stage the coil needs to generate a magnetic field, which safely opens the valve.

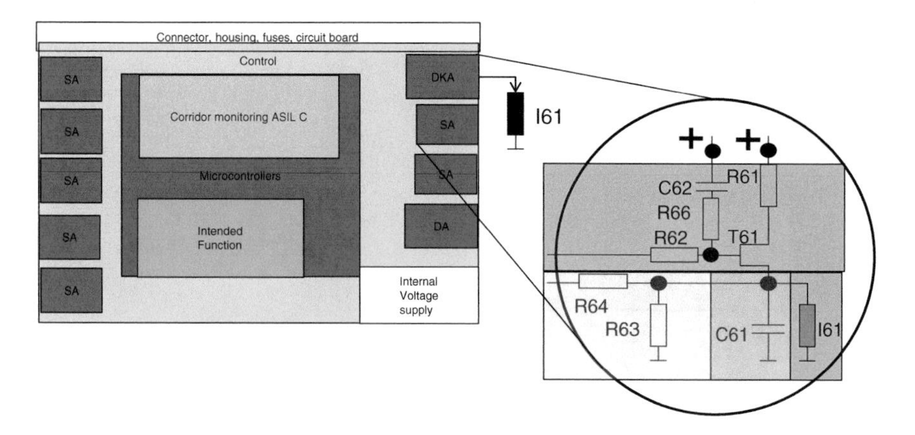

Fig. 4.74 Defining of E/E hardware functional groups derived from system decomposition

x.2 The current in the valve coil needs to be read back in microcontroller through the driver stage as analog value.

What is the benefit of or which improvement for safety do we get in this example by further breaking down the requirement in natural language? For the design, some calculations would ensure that adequate capacitors, resistors and transistors could be chosen. However, those components would be correctly chosen simply by the derivation of the requirements. In principle, "trial and error" determines an useful combination, which enables us to power-optimized get to a suitable realization according to the life span requirements.

We could get requirements, architecture (i.e. behavior, structure), constraints, analysis results (failure descriptions, failure assessment (severity-rating in FMEA)) and design (drawings, geometry guidelines) from the system development (see Fig. 4.75). According to the example the requirements, which are now allocated to the EE components, will be mapped to the system architecture. This means that the system architecture already provides the structure for the requirements, which now need to be broken down within the electronics components. This can already imply the first iteration but through an accompanied safety analysis it would be ensured that the malfunctions and their structures are consistently maintained. Based on the architecture analysis we can already make solidified statements whether the architecture derivations are consistent, complete and so also transparent. With this result we can now verify the requirements, meaning, we can test, whether there are sufficient requirements for all inputs of elements (information and configuration inputs), for all outputs and for all permissible input–output relations. The next step is to derive the design from the system. At this point requirements, architecture and analysis results should be available. At this point it is especially important that confirmed information, assumptions and unconfirmed or non-secured information are communicated to the designer so that they can assess their design-decision or what options are possible for variants. Besides these horizontal information, designer also receive information from the system (design limitations, geometry), which indicate limits, within which design decisions need to be made. With the derived design decisions we can now move on to the verification. In this context, all horizontal information is questioned, whereas especially the results of the analyses should be confirmed, for example through test. The following verification methods could be considered:

Positive tests (requirement based tests): Since through the analysis the completeness of the requirement specification referring the derived structure can already be confirmed, in the verification we can now test the correct implementation of the

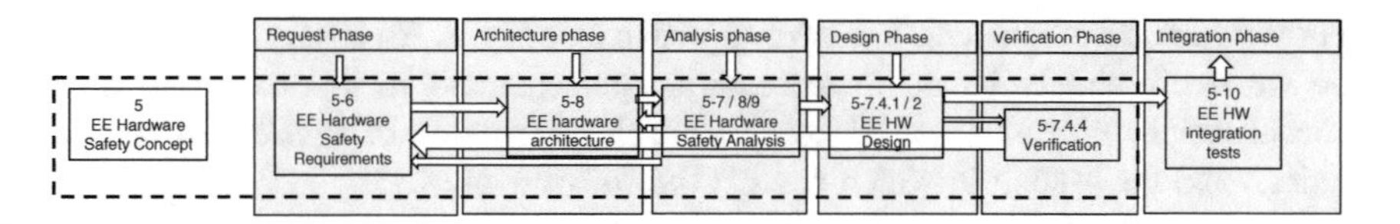

Fig. 4.75 Information flow and phases of activities in product development on hardware level

requirement in the design. Since all relevant parameter, which result from the technical elements and the permissible interactions, are known in the design specification, we can now also confirm the correct implementation under consideration of the design guidelines. The Design-FMEA is recommended in almost all quality standards for the design verification. In this context we question the design characteristics in their combination with and against their environment. In the automobile industry we refer to this as design verification (fulfillment of requirements) or design validation (fulfillment of requirements derived from upper levels (also customer or higher level requirements)). Validation is also often seen, that the question should be answered, if the requirement are correct. Fulfill the requirements their higher level demands, they are somehow correct. Often only the abbreviation "**DV**" is used for both **Design Verification** and **Design Validation**.

In addition to that there is also **"PV" (product verification, product validation)**, which should confirm that the lifespan requirements are met also for the tolerances of the production of supplied components. Design-FMEA formally questioned which error sequences occur if a characteristic is deviate from the specified range. How such faults propagate into the upper levels up-to a possible violation of safety goals can be assessed from the analyses and the architectures of the higher levels.

Negative tests (failure injections, limit tests, tolerance chain tests, stress tests (including EMC)) mainly show the robustness of components. The failure injection shows the correct function of the safety mechanisms, the correct assumption of the error propagation, the sufficient robustness, the behavior at inadmissible configurations and the compliance of functional and technical limitations.

This allocation to negative and positive tests should be seen as absolute, especially EMC specialists will also test the adherence to permitted values for a correct design and tolerance chains will also be positively assessed through given budgets. If a test is planed with the actually used technical elements or whether a calculation will be performed based on models, is a decision of the verification planning (test planning). If the verification is logically reasoned through models or calculations, expensive test setups can be avoided. In this case often a combination is chosen, since without tests it cannot be shown that the models or the calculations actually match the requirements for the realization. Tests to confirm the correctness of models general could be finding in literature as "model validation".

4.7.3 Software Development

For software development a V-model approach (see Fig. 4.76) is very common. But do we really simply break down requirements or aren't in this case also other considerations necessary, which result from the interaction of environmental conditions and the system in which context the software is used?

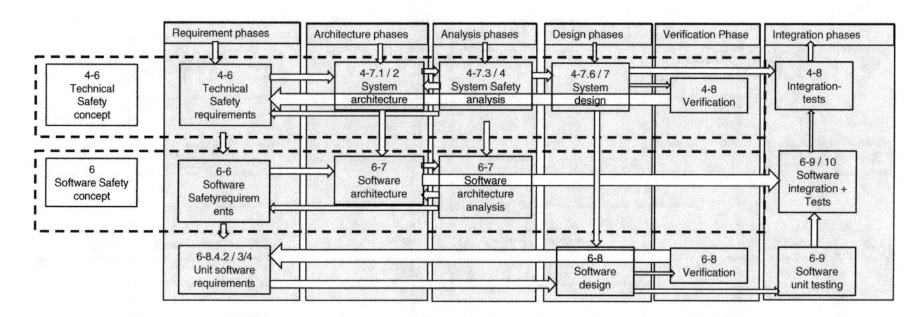

Fig. 4.76 Information flow and phases of activities in product development on software level and scope for software safety concept (SSC)

Tools such as the compiler, test tools, editors and the chosen programming language (including limitations for its use) will be framework conditions, which also influence the software development process.

The requirements for the software components are not directly derived from the functional derivations of the safety requirements, allocated from the system, but primarily from the software architecture draft, which result from all requirements and constraints, not only functional requirements.

Functional limitations and design limitations need to be derived from the environmental conditions and the design decisions of the upper abstraction levels. All these non-functional requirements mostly come from the draft architecture rather than the derivations of the requirements, which then need to be analyzed and verified referring to the requirements. This is why, in contrast to ISO 26262, the requirement and architecture development will in this context also be considered as software safety concept (see Fig. 4.76). For the software and microcontrollers, the following questions will be even more sensible: What is the difference between functional elements and technical elements? Do we describe ALU (arithmetic logical unit) of the microcontroller as functional or technical element? Probably almost all elements of the microcontroller are described as logical elements and we only need to discuss the degree of details of the elements and which characteristics actually need to be specified and described. This also applies for all elements and the software itself, since these elements are only describable through their functional behavior. If we consider a software unit as realized, we then consider it as a technical unit. It is true that ISO 26262 provides the hardware design specifications and method guidelines (from external sources) as further supportive information for the specification of software safety requirements (ISO 26262, part 6, Chap. 6.3.2) but only in ISO 26262, part 6, Chap. 7.4.5, in the software architecture design, we then find indications, which need to be considered for the static and dynamic aspects of the architecture. However, in this context there will be no sequential process, since we would need an architecture draft in order to even be able to derive software safety requirements. Also referring to the illustration of ISO 26262, part 4, figure B1 we can see the many influence factors, which only derive from the HSI (hardware–software interface) (Fig. 4.77).

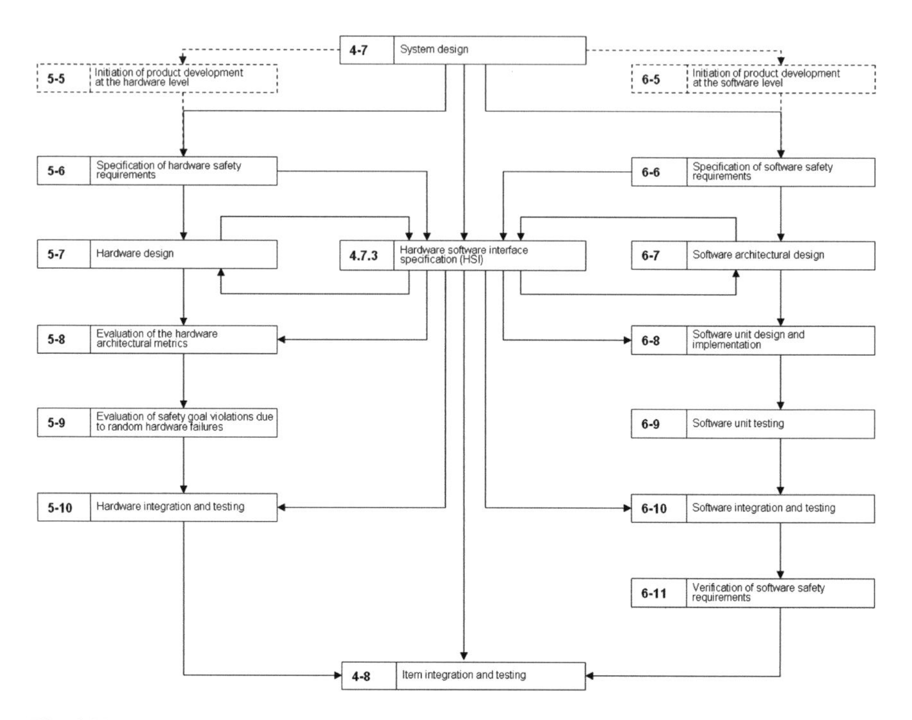

Fig. 4.77 Normative influence to the hardware software interface (HSI) (*Source* ISO 26262, part 4)

This means that besides the computer influences, coding guidelines, tools also the architecture decisions from the basic software are added, which need to be considered for the development of software safety requirements.

The challenge for the SW architecture analysis is to standardize the influence of the microcontroller to the SW architecture.

A standardized environment needs to be created for the application software, which cannot be influenced by technical impacts of the microcontroller. If we try to analyze all actually possible error influences of the microcontroller on the software with the SW architecture analysis, we will mainly need to test each characteristic of the microcontroller. This has a direct influence when we change a microcontroller but maybe also if the manufacturer of the microcontroller changes production technique or uses different materials. If the use of the software is intended for different computers, the development aim for the basic software should be to create a unique environment for the application software so that the user software cannot be safety-critically changed (Fig. 4.78).

However, this means that we will need very detailed information on the microcontroller, since we want to increase the performance through new computer architecture. Therefore, 'no influence' cannot be a development aim, but the safety mechanisms used in the application software need to be continuously effective even for the changed environment.

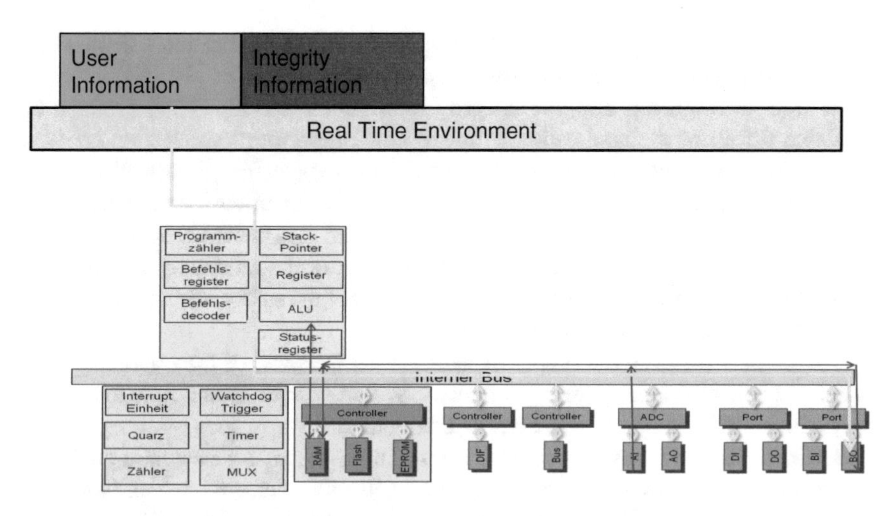

Fig. 4.78 Example: data flow at the hardware–software interface (HSI)

Basically there are the following errors, which can effect from the controller to the application software:

- Information could be false. Additionally, information is generated by mistake.
- Information could not be available in time ('stuck at', based on SW functions such as scheduling/program flow and on failures in the hardware of the microcontroller).

All other possible failure impacts by the microcontroller to the application software need to already be controlled by the basic software. However, it is a question of preferred software architecture, where the error types are safeguarded. It would be possible that the errors are controlled in the basic software. Especially data correction, control mechanism or implemented safety mechanism versus systematic errors from the peripheral, sensors and also from the microcontroller itself effectively implemented in the basic software would simplify the application software and related safety mechanism, If possibly the application software needs only safety mechanism against their own systematic faults or safety mechanism which are implemented in software but control the systematic failure on system level could simplify the needed architecture and related dataflow tremendously. Since safety goals are often also subjects to change, the safety mechanisms against systematic failures on system level should be implemented in an independent area.

In order to comply with the Autosar standard, the all data for the RTE (Real-time-environment) need as qualifier at the virtual function bus for any signal a safety qualifier or a diagnostic key. Typical to the Autosar the hardware (HAL) and microcontroller (MCAL) abstraction layer could be considered. An additional system- or sensor abstraction layer could be introduced, so that the application software already gets physical data according to the functional system design, so that the input data and the functions in the application become traceable and more transparent. Especially software safety mechanism in the application software and also their degradations could so easier map to system level.

In order to fulfill this aim, the integrity information (see Fig. 4.79) needs to be available for all of the application software in addition to all application relevant information. This integrity information provides corresponding information about degree of correctness and up to the validity of the information for the application software. How much in depth this diagnosis information is provided depends on the highest ASIL. For the information flow to the actuator the internal diagnoses need to be provided within the application software so that the safety relevant function at the actuator could be initiated (Fig. 4.80).

The outputs (or even the output register and so the pins) of the microcontroller could be controlled by monitoring functions, so that safety relevant functions allow only actions in case timely and correct data output,

The necessary diagnoses to activate certain information at the actuator are based on the integrity information. The diagnoses of the SW elements, which generated the output information, belong with the internal safety mechanisms, through to the integrity information, which transfer the information to the actuator. This integrity information should be processed independently from the application information. This can happen for the application data, diagnosis data and safety mechanisms by electronic signatures, pattern or encryption. Coded processing is not necessary in general even in case of ASIL D, the degree of asymmetry between intended function in the application software and in the safety software have to be assured.

For the application software it is sufficient to define the data formats for the application software. The hardware–software-interface (HSI) allocated to the basic software. Therefore, the software architecture analysis is limited to possible systematic failures, which are only reduced to the possible systematic failure of the application software. If in a sufficiently independent horizontal abstraction layer

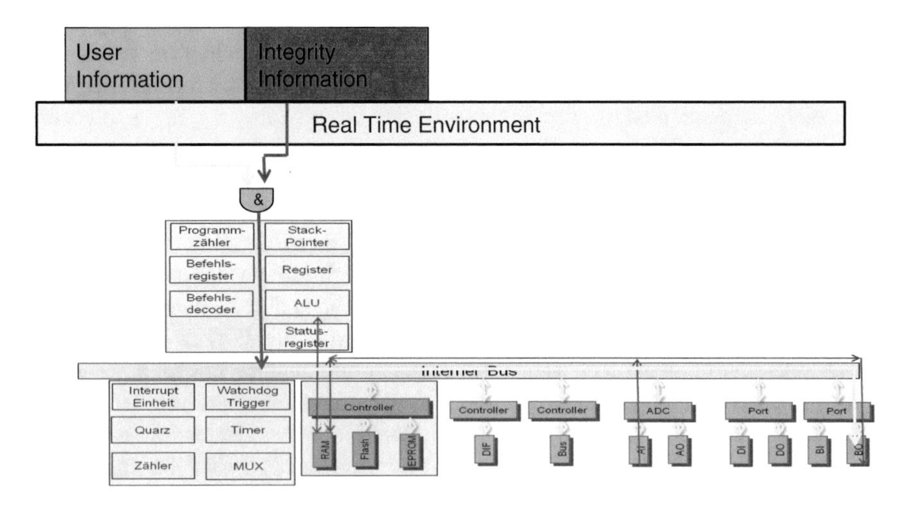

Fig. 4.79 Safety-related dataflow of safety software in addition to application software

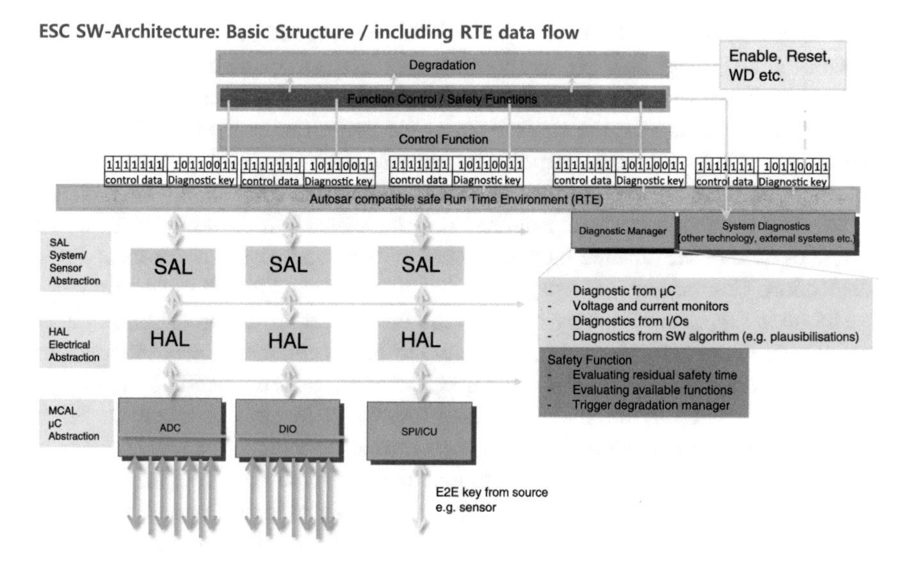

Fig. 4.80 SW-architecture based on Autosar principles with 3 (multi)-layer safety architecture

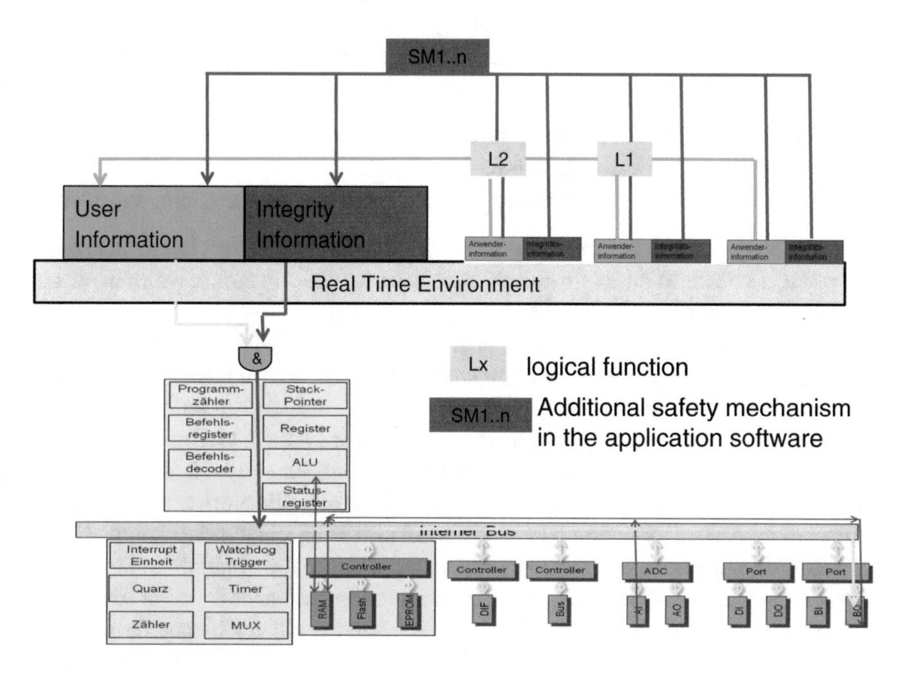

Fig. 4.81 SW-architecture: example: data flow at the hardware–software-interface (HSI) and interface via RTE

(such as for EGAS already at the system level) corresponding redundant function monitoring and diagnoses are introduced, that control systematic failure could derive from independence constraints on system level. If it is useful in the individual case, especially for the availability of the system or such architecture could lead to higher fault tolerance, depends on other factors.

Generally, there are even further sources of risk within software's, which cannot be covered by this approach. Routines can be called within software's, which operate outside of the covered hardware areas and can cause runtime errors and data falsification. Consequently the degree of independence is difficult to determine.

The operating functions may be up to the core codes of the processing unit, or resulting compiler settings or functions have to be specified sufficient independent as part of coding guidelines. If for example, a cache is not sufficiently controlled, it shouldn't be used for the safety relevant functions. In case of adequate safety mechanism in the basic software, in microcontroller specific software segments or even in the hardware, such implementations could reduce the effort for the application software. The coverage tests, which are required in the software design level in ISO 26262, wouldn't prevent or reveal such potential risks. We could only test known failure scenarios by integration tests and adequate fault injections.

This is why we would need to continue to implement function monitoring (compare Fig. 4.81) for the higher ASIL also through the safety relevant software functions. In this case there are multiple approaches, for which the function monitoring is actually effective. The function monitoring can safeguard the following three functional groups:

- the software functions of the application software
- the application software and the basic software
- the entire embedded software and the data interfaces to the microcontroller

It will be difficult to use mixed architecture in a distributed development, since at this point it comes to an explosion of interfaces for the function development and the failure analysis.

If the interfaces are determined in a way that the entire hardware is safely controlled below the RTE and the application data and their integrity identification is provided at the RTE for the application software, a distinct interface occurs.

If the basic software and eventually also the hardware integrity measures are safeguarded by function monitoring, the system can become more fault tolerant, but the safety analysis becomes highly complex. To really illustrate a double failure control to an ASIL D function will be extremely difficult, maybe even impossible to the enormous number of error combinations. If there are multiple safety goals, which require a failure reaction in different directions (a value too high or too low violates different safety goals), planning degradation will no longer be possible. Since switching-off will only be possible in the case of faults, the benefit of fault tolerances will limit the reliability of the system. Balancing of safety, availability, reliability with performance could become a huge challenge for a software architect. The designer of the software could most likely only limit or avoid the worst-cases.

References

1. ISO 26262 (2011): Road vehicles – Functional safety. International Organization for Standardization, Geneva, Switzerland.

ISO 26262, Part3, Clause 7 81

ISO 26262, Part 3, appendix B: 87

ISO 26262, Part3 94

ISO 26262, Part4, Clause6 104

ISO 26262, Part 9, clause 8: 120

ISO 26262, Part 6, Clause 7.1 123

ISO 26262, Part 6, clause 7: 123

ISO 26262, Part 6, clause 7 123

ISO 26262, Part 5, clause 8: 143

ISO 26262, Part 5, Clause 8.2: 144

ISO 26262, Part 5, Clause 9: 155

ISO 26262, part 5, appendix C: 155

ISO 26262, Part 5, Annex C 147

ISO 26262, Part 5, Annex C: 147

ISO 26262, Part 5, clause 8.4.7: 151

ISO 26262, Part 5, Clause 9.4.2.1 154

ISO 26262, Part 5, Annex F 156

ISO 26262, Part 5, Annex F, Example 1 158

ISO 26262, Part 5, Annex F, Example 2 158

ISO 26262, Part 1, Clause 1.22 161

ISO 26262, Part 1, Clause 1.13: 162

ISO 26262, Part 1, Clause 1.14: 162

ISO 26262, Part 9, Clause 7.1.1 163

ISO 26262, Part 9, Clause 7.1.2: 163

ISO26262, Part 4, Clause 7.4.2.4 164

ISO 26262, Part 9, Clause 7.44: 165

ISO 26262, Part 9, Clause 7.4.7: 166

2. Marcus Abele, Modeling and assessment of highly reliable energy and vehicle electric system architecture for safety relevant consumers in vehicles, 2008.
3. *Deep Medhi. Proceedings of 7th International Workshop on the Design of Reliable Communication Networks (DRCN 2009), Washington, DC, October 2009.*
4. VDA (1996), Volume 4 FMEA, Frankfurt.
5. SAE J2980, Considerations for ISO 26262, ASIL Hazard Classification, Prop Draft F: 2011ff.

Chapter 5
System Engineering in the Product Development

The general approach of all system engineering standards is the realization of products based on their specification. In the context of a V-model it is the bottom of the V; quasi the end of the descending branch and the beginning of the ascending branch. The elements to be realized are technical elements of hardware or software. For hardware either electronic or mechanical a production process is necessary, for software mainly the tool-chain built also a kind of realization or production process to effectively build the software. If previous analyzes and verifications were all successful, correct and sufficient, the integrated elements should work as specified and all observable characteristics should meet the required expectations with respect to performance, all other quality factors and intended functionalities should be achieved. When to perform safety activities and what kind of safety activities are needed results from these requirements, which had been developed during the descending branch of the V-cycle. The realization of components or products should be considered as the bottom of the V-model and any integration from the smallest part or unit should follow the requirements addressed to the ascending branch of the V-cycle.

5.1 Product Realization

A design description is not yet a realized product. Placing components on printed circuit boards, linking together different mechanical components, generating, realizing or integrating various software elements or integrating or linking together components based on different technologies; all these activities will influence the correct technical behavior and functionality of the product. Especially the realization and integration of elements or components of different technologies are only partially addressed in ISO 26262. Most safety standards do not even impose any requirements on the product realization itself. Also, how realization should be performed based on specification is often left to free interpretation. Referring to the

H.-L. Ross, *Functional Safety for Road Vehicles*,
DOI 10.1007/978-3-319-33361-8_5

verification of the design, ISO 26262 requires that the product or the component, among other things, be completely and consistently specified. This can be achieved through requirements, architecture (block diagrams, behavior diagrams, models etc.) or design documents (design specifications, parts lists, drawings etc.).

5.1.1 Product Design for Development

Nowadays, almost all development standards assume that a product is developed based on specifications. However, given that, it is important to ensure that specifications consist of at least two work products, namely the requirement and design specification. In the classical mechanical development, the construction drawing came at the end of the development process. After several tests (verification) and the completion of the production facilities it received its approval for production. If we assume that the design developed based on the requirements and was also verified and validated, we can still use this classical approach for software based systems or products. The process up until the development can be described as descending branch of one or more V models or through spirals (model cycle in the automobile industry) or waterfall models, which are used as basis for the product, project or process maturity assessment. If the product or process isn't completely, consistently and transparently implemented all the way through to the product development, systematical failures in the product will occur. How such project or process failure will influence the characteristics of a product can generally not be assessed or predicted. This is why the ascending branch of the V model is used in ISO 26262 to get to a product assessment through systematic integration and further tests for the validation and verification. The development of hardware components, no matter whether they are mechanical, hydraulic, electrical or electronic components, strongly depends in the production resources and their maturity referring to serial production. The software will more likely be developed in a laboratory environment. In analogy to hardware products or components, the development of software elements and their fulfillment of non-functional requirements such as security, quality and reliability etc. could be considered as a so-called software factory. The necessary activities to ensure correct software elements are basically very similar.

5.1.2 Mechanics

The characteristics (features, capabilities, properties) were identified in the context of the requirement development. Technical elements can now be developed through standard parts (screws, nuts, plugs etc.) or mechanical parts, intended for the application. In any case, it is important to secure identified important characteristics within the complete specified environment. This means that the thread can be seen as relevant for the bolt and the nut but other characteristics will be also important. The lack of a certain tool such as a matching wrench for the screw can be relevant in regards to safety.

However, if the tightening torque for the screw is identified as an important characteristic, the choice of tools or the monitoring of the work with the tool may become a safety activity. The production (for example the development of the link through screwing in the screw) needs to be monitored primarily to show that the requirements are developable (often for the first prototypes) and also correctly implemented. "Special characteristics" are often determined for safety relevant characteristics of the top products. Those are often tested in the serial production for each individual product and the test results are filed in data archives.

Traditionally we will find four sample phases in the classical development of the automobile industry. The following aims are considered for mechanical parts for these samples:

- A-sample: shows the geometry (form), fit and function fulfillment
- B-sample: ensures the functionality and endurance on the test bench and for the prototype
- C-sample: ensure the functionality, endurance, compatibility and integration ability in the target application (engine, vehicle)
- D-sample: (initial sample) released sample for serial production

Generally development processes are also described in phases, which are oriented on historical sample phases.

The A-sample (concept phase) is often already considered in very early phases of the development and shall be provided along with the offer of the supplier.

The B-sample (design phase) is often the sample with which the design is verified and validated (DV). This means that all characteristics should already be secured. As a result, all requirements should be verified and available and the architecture and the design also need to be analyzed and verified. Production concepts should be detailed enough, that Process-FMEAs and safety-related activities during production can be identified. All safety features or safety mechanisms and related characteristics shall be correctly implemented, tested and verified.

The C-sample (industrialization phase) should be able to be implemented and compatible in the target environment, for example, tolerances of supplier parts or accepted tolerances of product parts. Resulting tolerance or tolerance-chains from the production of assembled parts should be secured within specified limits. Therefore, it is often required that the C-samples are already produced on series development machines or at least by using tools for series production.

For new products, the production machines and related tooling need not be aligned according to the target production process. If not, the machine interfaces and the entire production chain could not be qualified. The production process interfaces lead to further no acceptable product tolerances. The necessary machinery and process capability could not be shown.

D-samples (ready for series production phase), should be produced in a qualified series production process including all machines and test facilities as defined in the control plan and during PPAP (Production Part Approval Process). During "Run-at-Rate" a defined number of samples should be produced so that the quality characteristics and the performance of the production process can be assured also during series development conditions. The produced sample is then seen as the basis for the customer (e.g. the vehicle manufacturer) for the release for serial production.

Consequently, all product characteristics identified as safety-related have to already be assured during the B-sample phase. In today's development cycles it is nearly impossible to identify all safety related characteristics for electronic and software elements. However, requirements and characteristics for electronics and software can be deriving from the characteristics of mechanical components. This approach is also highly recommended for mechanical parts of electronic and electrical elements, such as connectors, housing, printed circuit boards etc. the approach is highly recommended.

5.1.3 Electronics

Basically, the electronics is developed much in the same way as mechanics, based on design documents and also produced within different sample iterations, which makes the description of the sample phases comparable. As already described for mechanics, a lot of design parameters for electronics depend on mechanical parts such as the printed circuit board, plugs or housing. This means that theses tolerances or tolerable discrepancies are the basis for the design of the electronics. Especially for geometric characteristics there are a lot of characteristics to be considered.

The housing has to be constructed in a way that it fits in the vehicle, provides protection against humidity and dirt ensures that cables can be fixed, fulfills the EMC requirements and allows the arising heat dissipates. It is now possible for the development of the electronics to illustrate a specific separation to mechanics. This is why testing the samples and serial parts will be one of the essential safety activities.

5.1.4 Software

In opposite to hardware, software is not produced in a facility but adapted within the development. Where software development starts is interpreted differently in various standards. ISO 26262 assumes that the SW unit is the smallest element in the architecture and that for example the C-file represents the smallest unit for the code generator or compiler. The binary code could after linking transferred (flashed) into the (flash) memory for (of) the microcontroller. Random failures can occur in this process so that the implemented code does not correspond to the source code.

There are many requirements for safe compilers, but the entire logistics of sw-code linking and any configuration required for compiler, flashing etc. could lead to undetectable systematic errors. Furthermore, for model based software development, a binary code is often already generated through a code generator in a software module, which consists of several SW units. At this point the question arises how SW units interact and whether the intersections between the SW units may be negatively influenced through the computer environment (Fig. 5.1).

Not only the code-generation or the compiling and flash processes could lead to systematic errors but also the integration of libraries (headers, static and dynamic libraries) could be a source or cause of systematic errors.

The tools that support these activities are not always qualified to sufficiently prevent possible errors. The realization for a tool qualification was the main motivation for part 8, Chap. 12 "SW-Tool Qualification" of ISO 26262. Later it had been identified that almost all tools could affect the safety of software-based products. The main idea of a V-model process could also support those tool influences during the software development. If we go all the way down to the SW units and completely test those corresponding SW units through the coverage test, a negative tool influence can be excluded.

In case of a systematic integration processes negative tool influences can be identified during integration or at least during the verification of the integration. With this approach even negative hidden systematic hardware impacts, e.g. from functional units of the microcontroller, can be identified. Through analyzing positive and negative results from integration tests, it is even possible to trace the cause of errors to tool influences or other systematic errors, which can only be controlled by implemented mechanisms.

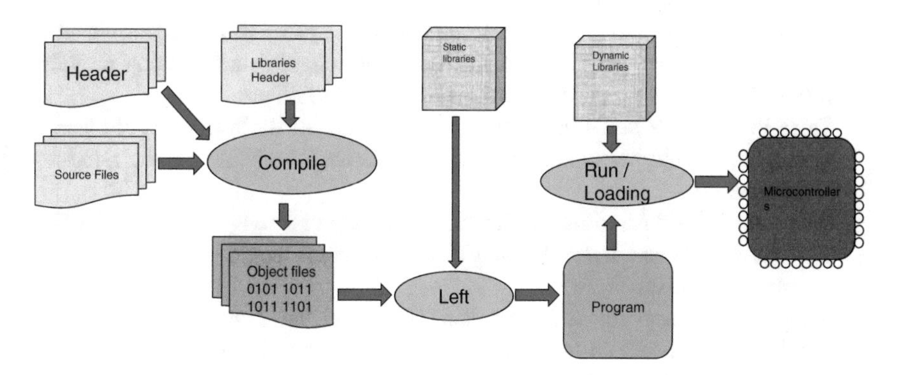

Fig. 5.1 Principle of tool chain for C-coding

In practice, there are some gaps on the hardware level because of, for example, incorrect handling of tools or insufficient test possibilities.

For ASIL C and D software, ISO 26262, part 6, Table 4 requires additionally implemented plausibility checks and control flow monitoring; in case of ASIL D also diversity (ASIL D) in software as well as data and control flow analyses are required (see part 6, Table 6). Especially Table 4 requires implemented safety mechanism (e.g. redundancy), which should assure compensation of such systematic errors during software development. Even if safety mechanisms show sufficient effectiveness on the software architecture and design level, they should be verified after the realization and implementation.

Implemented code could influenced by systematic faults not only during design and compilation, it could be affected by any activity related to the logistic or handling of the embedded software.

5.2 Functional Safety and Timing Constraints

5.2.1 Safety Aspects of Fault-Reaction-Time-Interval

Automotive safety systems consider real-time aspects as “real-time constraint”, for example operational deadlines from event to system response. Similar to requirements deriving from an ASIL-Decomposition such as sufficient freedom of interference or independence, system requirements give no guidance on how to design, or implement timing constraints. During the verification, especially during stress tests, fault-injections, over-limit or worst-case testing such constraints could be analyzed and potential violations identified. For example real-time programs must guarantee response within strict time constraints, often referred to as “deadlines”.

The following aspects could be considered in safety-related automotive systems to be processed within a defined time-interval:

- Typical fail-safe systems require a shut down or a de-energized state after a safety-relevant fault occurs
- Calculations, functions, embedded simulations or processes must terminate so that further actions can be initiated. In case of correct data not being provided within the defined time-interval, the system could fail.
- Close loop control and control data (e.g. influence of death-time) could lead to wrong timing behavior resulting in control interventions being too slow, too fast or delayed etc.
- Data must be compared when filtered, in the case of wrong timing for the filter (e.g. safety-relevant events) safety relevant effects could occur such as an overshoot or noise.
- In the case of comparing 2 drivers data sets (e.g. signals from a sensor), the data could have different run times (age of the signal) referring to the measurement

and processing of the data (yaw rate needs 200 ms to the processor and the steering angle could be provided each 10 ms, so that the driver already changed the driving direction).

- Communication systems should provide data. Without a safe time-stamp, which considers the individual age of the data, or in case of a sequence and event recording or detection, the first event could not be determined.
- Data interfaces such as a virtual function bus should be continuously up-dated, in case of differing age of data in a common runtime-environment; the data cannot be used for further safety-relevant actions or commands etc.

Some of the functions listed are not always typical safety-related functions. Also, in case of a precise control or just for data or event-synchronizations, time-constraints must be considered. Especially in embedded systems several time-constraints could be required in different contexts within a single micro-controller and multi-tasking principles must be applied. In case of multi-core applications, such requirements are relevant at least in order to manage common resources, such as peripheral elements, packaging, power-supply etc.

5.2.2 Safety Aspects and Real-Time Systems

Real-time is not always an aspect of short time intervals; sometimes such an interval allows seconds or even minutes. For example, switching-on the vehicle light during the start of vehicle will be a matter of seconds. The driver should realize when the light switches on, and the correct functioning light should be available before starting or driving. Even in a case of switching it on during driving, a delayed reaction for a few seconds would not lead to a safety-relevant impact because in most areas of the world it does not get dark that quickly in the evening. The acts of switching-off and switching-on lights are safety-related, but the time from the demand to the illumination requires only about 1 s. An exception could be 'high-beam'; it should be changed to driving light within less than 1 s, to avoid glaring of on-coming-traffic.

Very often we can find the following definition of real time systems:

"A system is considered to be a real-time-system, if the correctness of an operation depends not only on the logical correctness; it addresses also the time in which it is performed."

Vehicle real time systems in the context of functional safety could be classified by the consequence of missing a deadline:

- Hard—missing a deadline is a total system failure. In case of a safety-related real time system, missing a deadline leads to a violation of a safety goal. Electronic steering and brake systems of vehicles could be considered within this category.

- Firm—missing a deadline is tolerable, may degrade the system quality or performance but does not lead to the violation of relevant safety requirements. A motor management system could be considered as a typical example.
- Soft—deadlines are specified for the system but the tolerances for the deadlines are so high that safety impacts or even violations of safety requirements or goals could not be credibly argued. A light system of a vehicle could be an example, because in general drivers themselves can control missing deadlines.

It could not be considered, that chassis systems considered as hard real-time, power-train systems as firm real-time and interior system are soft real time systems. ISO 26262 defines the fault-tolerant-time-interval as the basic criterion to specify safety-related timing-specific requirements. This interval defines a period of time, within faults should be controlled by the system (see also Chap. 3 of this book). The consequences of a missing deadline can only be evaluated by the consequences of the fault-error propagation; in the latter, consequences of the possible hazard due to resulting malfunctions.

The example with the light system shows that also very often the average time of fulfilling a deadline and the distance to the worst case time could be seen as a criterion for a soft or hard real-time system. The time-distance shows how probable or frequent the potential is to miss a given deadline (Figs. 5.2 and 5.3).

If a real-time deadline is allocated to a safety-related function the deadlines must be met, regardless of the system load. This heavily impacts the software architecture, design, realization, and implementation.

The following effects can be considered to distinguish between hard, firm, and soft real-time applications:

- Consequences of breaking deadlines
- Relation of average and worst-case execution time
- Tolerance time interval leading to failure of the system.

All 3 effects could lead to different design criteria for the implementation of software.

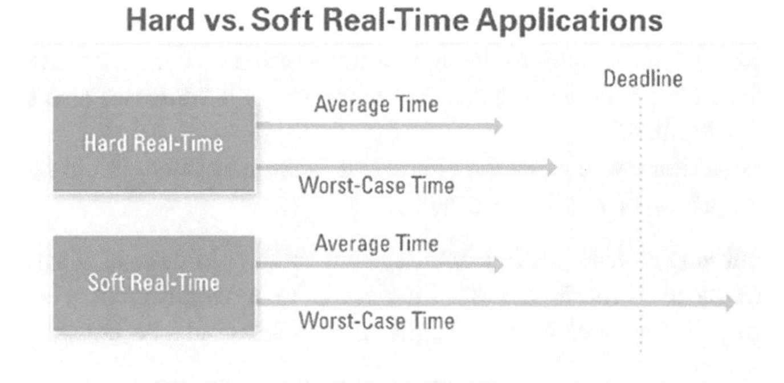

Fig. 5.2 Hard versus soft real time systems in the context of average execution time and worst-case execution time related to a given deadline

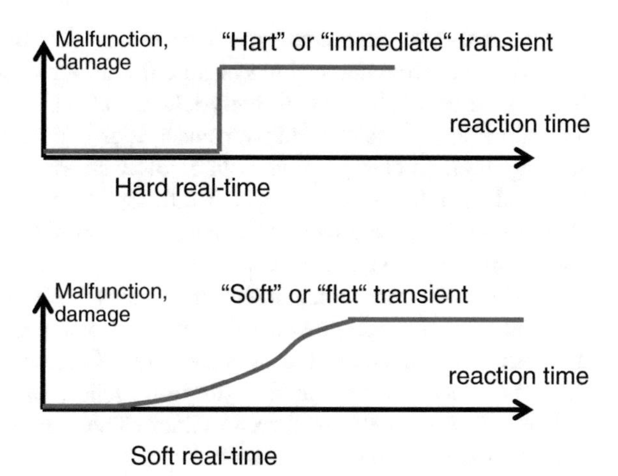

Fig. 5.3 Hard versus soft real time aspects from typical PC-Applications

5.2.3 Timing and Determinism

In real-time-systems determinism is one of the key-features for the design. A system or a communication could be call deterministic if the maximum response time could be predicted. It means that no effect whether systematic errors nor random hardware faults or other effects could violate these maximum response time, data refresh time or other periodic effects. The periodicity, data refresh time or response time does not mean in case a short time, but similar to the relation between average execution time to worst-case execution time for hard real-time-systems, the predictability is the key criterion. It could be predictably fast or predictably slow. Serial data communication allows very often deterministic behavior. For example standard **Ethernet** is not real-time capable, because it is not deterministic. The lack determinism based in the fact that the traditional Carrier Sense Multiple Access/Collision Detection (CSMA/CD) principle detects collisions, but they are not avoided. Due to point-to-point communications, full duplex transmission with switched Ethernet could reduce significantly the probability of collisions. Repeating of data transmissions could also improve the availability. Systems which relay on a safe availability of the communication need independent redundant implemented communication lines. If Ethernet is design redundantly a logical ring structure could improve the safe availability of communication for such communication systems.

Determinism of Ethernet is well discussed, but also determinism of algorithms sensors or other elements in a system could be required. Typical examples are; a sensor has to provide each 5 ms a new set of data, a simulation model in an embedded system shall provide each 50 ms up-dated values, a position sensor in brushless motor shall continuously provide in a fixed time frame the motor position.

A steering angle of a vehicle could be always transmitted only in case of changes. During parking of a car or even during straight forward driving no data must be transmitted. Using a steering angle sensor in a safety related system could lead to real-time requirements and necessary deterministic behavior of control algorithms and communication systems. If a driver steering intervention could not be detected or just too late, a safety-relevant effect or failure could be considered. Therefor it today's systems implements a deterministic periodic transmission of the steering angle to chassis systems like power steering or electronic brake systems. By exciding of the maximum specified transmission time the value from the sensor shall be inhibited or a redundant model or a virtual sensor could provide correct data within the maximum time interval.

Isochronous data communication systems like **Flexray** operate on a precisely periodic basis, for example on a time interval of 1 ms. Bus cycle jitter is very low and generally far below the nominal 1 ms. Caution, bit transmission jitter is a different effect. Bit transmission jitter or cyclic transmission jitter could lead to different functional effects, therefore jitter effects should be considered as different characteristics of fault or error modes.

Synchronized communication and control algorithm are coordinated in a way that input data are taken, then communicated, control output computed, then communicated, and lastly actuated.

Many communication systems are "free-running". Mechanisms like time monitors, message counter provide sufficient evidence that soft deadlines wouldn't violated. In case of hard real-time systems the impact of exceeding a deadline could only lead to a failure of the system. In case of chassis systems a brake demand, a steering intervention or a curve of the road would not detected by the system within a required time-interval, which could lead to a violation of a safety goal or even worst to an accident.

The synchronization of communication systems can be made in different ways. The following principles could be considered:

- By one time slot method, the synchronization can be derived from the cyclical log. The synchronization based on sending a synchronization signal which is cyclically received and evaluated by all network participants. To ensure a best possible synchronization, it is required that the signal in a fixed time interval with minimal timing variances will be sent and received.
- Due to increase of temporal precision and synchronicity especially Ethernet systems based on the principle of distributed clocks (see also IEC 61588, Precision Time Protocol (PTP)), which are synchronized with each other via appropriate telegrams. Distributed clocks provide an accurate time base that is independent of maturities and fluctuations on the communication medium. Since this time base but can make sure no determinism for data transmission, the messages always with enough lead time must be transferred so that they become the synchronization time to processing available. Real-time Ethernet distributed clocks are used protocols to reduce the jitter occurring in the cyclic transmission. The communication system is not deterministic, but data transmission could be controlled in a way that sender and receiver have a deterministic data exchange.

What principles are safer, more accurate or higher available based on the application, amount if data, size and complexity of communication systems etc., both principles should be analyzed by means of a scenario analysis, including relevant safety analysis.

5.2.4 Scheduling Aspects in Relation to Control-Flow and Data-Flow Monitoring

Real-time embedded system should be designed in a way, that the real-time aspects are covered as precise as necessary, and performance and timely responses should be achieved as good as possible.

This principles lead to a conflict by using microcontroller resources. Usage of available resources form the microcontroller have to be planned and adequate principles like periodization etc. have to be implemented.

There are 2 general principles how to manage resources:

- **Non-preemptive**: processor or peripheral elements resources are assigned for functions or processes etc. and they release it for other processes or functions by their own mechanism. The resources could not be used by other processes or functions unless they are released by the assigned element.
- **Preemptive**: Resource could be taken away and be used by other processes or functions and return after their execution.

The operating system could make 2 types of decision related to resources:

- Assignment: "what processes or functions get what resources" The challenge is, resources are not easily pre-emptible.
- **Scheduling**: "how long are processes or functions assigned to resources". When more resources are requested than can be granted immediately, in which order should they be serviced? How could one processor share to many processes; memory, ports, DMA usage etc. The challenge is, making resources pre-emptible.

Hard real-time-system, especially if no timely response lead to a violation of safety goals with higher ASIL, the system should base on a high (or fixed) prioritized round-robin-principle.

Mix-criticality Application in hard real-time systems

New vehicle control units need solutions for different ASIL within a single control unit. If one of the safety-related functions even requires time-constraints, the assurance of sufficient independence is difficult to demonstrate. There are two principles which allow different ASILs in one microcontroller:

- **Multi-tasking**
- **Multi-core**

Multi-tasking generally based on single core or single core lockstep controllers. Most approaches with single core lockstep have only the advantage, that random hardware faults of the core functions need no additional software safety mechanism, even for ASIL D. But single core lockstep could not avoid systematic faults or errors in the embedded software. On both cores of the single core lockstep operates the same identical software. Due to the need of separation of, or at least sufficient independence of software safety mechanism and the intended function of the system a separation on task level becomes necessary.

Running different tasks on a single core requires sharing of resources. For higher level ASIL, the dependent failure analysis becomes endless complex; even what resources are used depends on compiler settings and programming style. Even very good software coding guidelines have to be questioned in case of new instruction sets of the microcontroller core. Changing of compiler or even microcontroller could have a tremendous safety impact.

The following example considers a safety task which needs to be repeated each 1 ms. The millisecond is necessary to synchronize I/O-Data, especially data with no time-stamp, and to trigger the watchdog, or any other degradation mechanism in the system. 3 basic functions are considered which run on a microcontroller with only these tasks for 3, 5 and 7 ms (net task time). The pure addition of the run time leads to 15 ms. In case of an interrupt each millisecond for the higher level safety task, the processing time double to 30 ms (Fig. 5.4).

Since the task itself does not have any time-constraints, such solutions are acceptable.

Typical active safety functions such as for chassis control, where for example typical loops for the application software are required, higher scheduling applications are required. If the safety task requires a higher level ASIL and the application a lower level ASIL or even only QM (e.g. legacy code), timing constraints violate

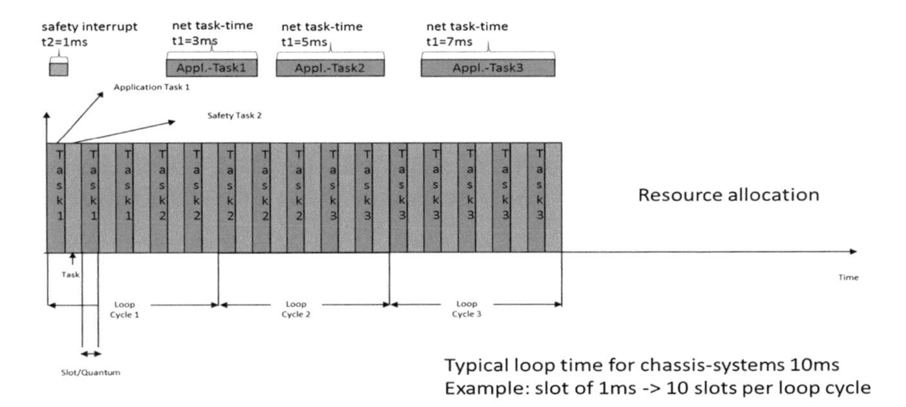

Fig. 5.4 Scheduling diagram for mixed criticality based on multi-tasking solutions

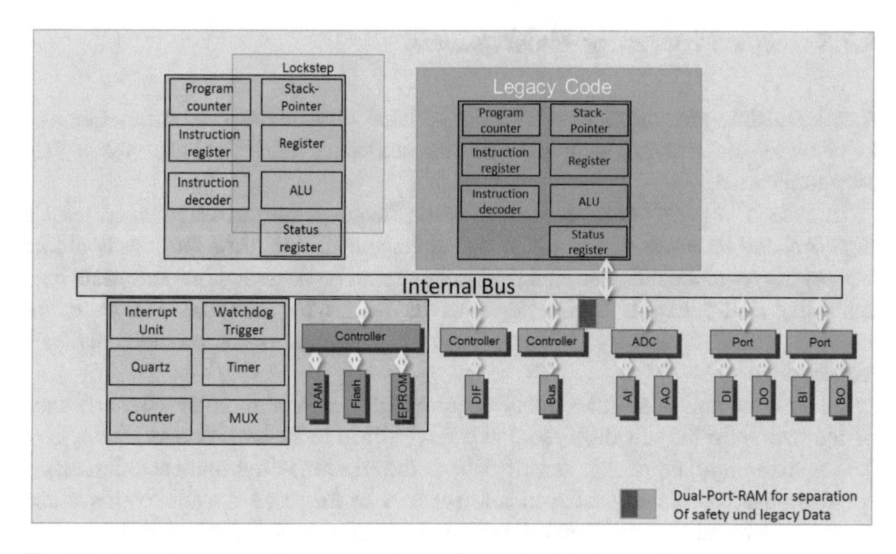

Fig. 5.5 Asynchronous multi-core controller with mixed criticality applications

with context switch between the tasks with different ASIL. Depending on the complexity of the application function and the ability of the microcontroller and operating system, such context switches lead enormous time exposure for the time. There are interrupt routines which run in an order of magnitude of nanoseconds, but the lowest time for the context switch is in an order of microseconds. That leads to very limited time slots for the application tasks.

In today's multicore solution, the microcontrollers offer an assignment of the safety function to one core and using the second core for the lower ASIL application function. Since even here the lockstep solution offers no measures versus systematic faults or errors, a sufficient independent monitoring layer is needed also on the safety core (Fig. 5.5).

By using multi-cores impacts due to dependent faults or errors are a challenge, since the kind of microcontroller, the compiler and the operating system could decide on which common resources are used within the microcontroller (core functions uses different resources). Changing of microcontroller, operating system and/or the compiler (or just settings) are nearly impossible and if necessary highly safety critical.

In order to avoid interferences between the data exchange of peripheral elements and with common memory resources a dual- port-RAM or message passing principles could be used, which would be monitored by the higher level ASIL core. Such a RAM-interface would be handled like a communication interface. Any other external peripheral elements shall be controlled by the higher level ASIL core.

5.2.5 *Safe Processing Environment*

A **safe runtime-environment** is for safety-related applications need safety-qualifier to show if the relevant data provide the sufficient safety integrity for a safe user-application.

In case of any safety-related absolute or frequent time constraints, at least a safety-related time monitor needs to be implemented. This time monitor could be used to trigger a watchdog so that the system could be degraded in a safe state by a controller shut-down. If a shut-down of the controller isn't the safe state, for example in fail-operational systems such degradations don't provide any safe reactions.

In case of soft real-time safety requirements such a monitor could change periodization for the scheduler, so that safety-critical tasks get processed. In case of deterministic up-date of e.g. sensor data at the run-time environment information could be used to identify delayed information. In this case the application could provide adequate safety-related function, so that the given safety-requirements would not be violated.

In case of hard real-time safety requirements timing must be controlled. In many applications, it is a matter of the safety architecture, if the lower time interval always requires hard reactions of the controller. Even if the controlled deterministic timing for fail-operational functions lead in case of errors to violation of safety goals, a shut-down of the microcontroller doesn't provide any safe reaction to the system. In this case the operation of the microcontroller is the only possible solution. In airplane applications TMR-systems (Triple-Modular Redundancy) are compulsory. Those systems mainly do not react on diagnostics and comparisons; they are mainly based on a 2 out of 3 (2oo3) majority voting principles. If 2 of the 3 results are equal, they provide the input for the actuator or any other final element. To assure continuous operations, the systems do continue they operation, even in case of detected faults. So that the as defect identified element could run in a recovery mode (reset of a controller) and after recovery all 3 elements could continue to control the safety critical actuator. The challenge is that such a 2oo3 system is well synchronized. In case of time delays the voter gets too much unequal results, which usual does not lead to a shut-down but in degreased performance of the system.

Consequently a safe hard real-time environment, need the nominal input information (e.g. from sensors) and a qualifier providing integrity information about the input signal and in addition a second qualifier that provides the timing information. Such a qualifier could provide information that the signal is within the required time-interval, but also about their synchronization such as delayed or too early. In case of safety-related control of vehicle direction, even the time-stamp from the detecting sensor (the origin of the electrical signal) could be provided.

Such principles could also run on task level, so that 3 dissimilar tasks provide adequate safety-related control information and by 2oo3 voting the actuator (or any output) provides the control information. In a single microcontroller application, such software-based voting does not provide a solution against entire controller impacts such as lightning flashes which immediately kills the system. But what level of hardware redundancy is necessary for a system, is a matter of the system analysis of the entire item (the vehicle system within its road traffic environment), rather than a question of the software or microcontroller architecture.

Chapter 6
System Integration

Integration starts with the smallest elements and ends with the validation of the development targets. Electronic hardware could be considered to be ready after the placing of the components or parts on the printed circuit board and assembly of mechanical hardware such as connectors, housing, cooling devices, and harness etc. Software integration also starts with the smallest units according to the make files and liking until the entire embedded software could be integrated and flashed into the microcontroller. After the hardware-software-integration the further components, sub-system or system elements will be integrated according to the hierarchical structure as they had been specified in the descending branch of the V-cycle, so that the acceding branch of the V-cycle matches in the horizontal layer of abstraction. After their realization or the integration in lower level of abstraction, the individual technical elements should fit with their interfaces. If the interfaces don't fit, the specification of the interfaces or other systematic error leads to these mismatches. The error have to be corrected according to the change management process, so that in a further integration step the interfaces become consistent. Since, the different activities in the realization and integration of underlying layers of the product or of components may not be adequately safeguarded, a lot of safety characteristics cannot be really verified and approved until the integration of the elements. In theory we should not get any new information at this point, since any characteristics and performance data, which are required from the interaction of the elements, have already been subject to the safety analysis and verification in the different horizontal levels. Since the elements have already been tested in the different sample phases, the changes, which are not considered in the previous sample phases, are usually the biggest risk. According to ISO 26262 [1] there are at least the following three system integration levels for a complete software based systems:

- Vehicle integration
- Components integration (System integration)
- Integration of software into hardware

H.-L. Ross, *Functional Safety for Road Vehicles*,
DOI 10.1007/978-3-319-33361-8_6

The entire embedded software usually will also not be integrated in a single step approach. A multistep integration strategy would be usually considered. The following software elements (groups of elements, components) need to be considered for the integration:

- low level driver (MCAL, microcontroller abstraction layer)
- operating system
- scheduler (program sequence, controlling, monitoring, data flow monitoring)
- run time environment
- application software
- software components with different ASIL
- degradation matrix
- hardware or system abstraction layer
- communication interface, busses
- error memory
- diagnosis or event memory
- safety mechanism to control systematic faults in system, hardware and software

How and in what order these elements should be integrated rather depends mostly on organizational interfaces or availability of the elements than the technical aspects. Integration should be accompanied by continues verifications and adequate tests to confirm the fulfilment of the given requirements for the relevant horizontal level of abstraction.

6.1 Verifications and Tests

The target of verifications are to provide evidence about correctness, consistency as well as completeness and therefore the transparency of requirements and their traceability to functions, characteristics of functional, logical and technical elements becomes adequate for the various stakeholder of the product. If these criteria for all functional safety aspect are sufficiently fulfilled, we could provide a sufficient basis for the safety validation. Verifications should not only be done during the integration (ascending branch in the V model) but also play a major role during the product development (or during the requirements development in the descending branch of the V model). Whenever a work result should be used as the basis for further architecture decisions, a previous verification is recommended; otherwise the following work results are only as useful as the underlying requirements.

ISO 26262, Part 4, Chap. 5

5 General

5.2.1 The necessary activities during the development of a system are given in Fig. 2. After the initiation of product development and the specification of the technical safety requirements, the system design is performed. During system design the system architecture is established, the technical safety requirements are allocated to hardware and software, and, if applicable, on other technologies. In addition, the technical safety requirements are refined and requirements arising from the system architecture are added, including the hardware software interface (HSI). Depending on the complexity of the architecture the requirements for subsystems can be derived iteratively. After their development, the hardware and software elements are integrated and tested to form an item that is then integrated into a vehicle. Once integrated at the vehicle level, safety validation is performed to provide evidence of functional safety with respect to the safety goals

Figures 2 and 3 show how integration levels could be planned by adequate hierarchical structure in the requirement development phase of the descending branch of the V-cycle.

ISO 26262*, *Part 4, Fig. 2—Reference phase model for the development of a safety-related item

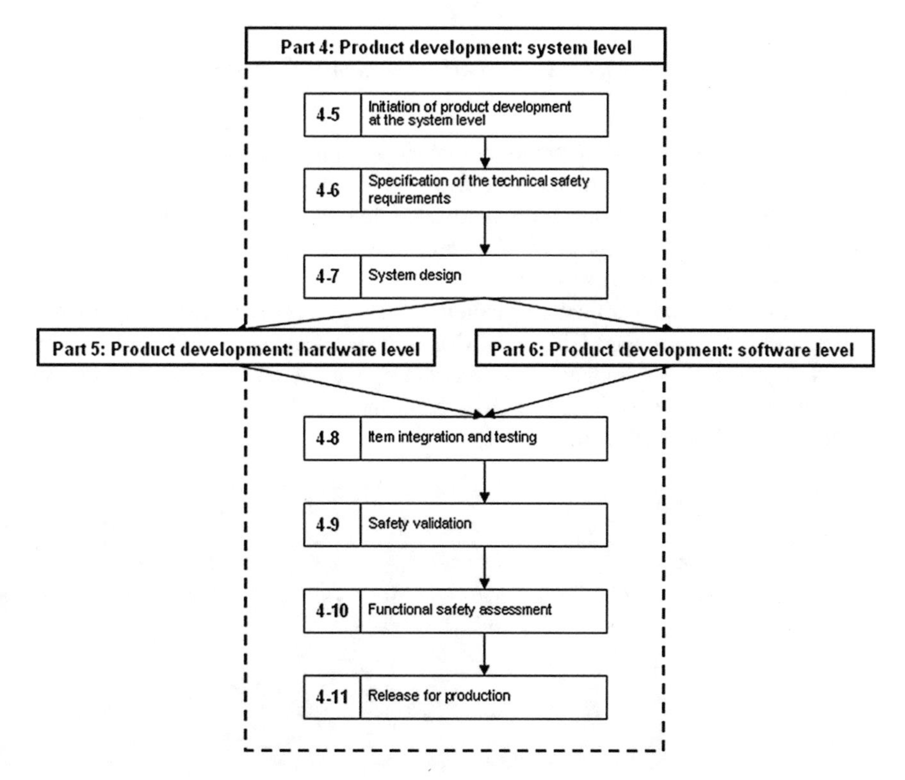

Figure from ISO 26262: Reference phase model for the development of a safety-related item (Source: ISO 26262, Part 4, Fig. 2)

ISO 26262*, *Part 4, Fig. 3—Example of a product development at the system level

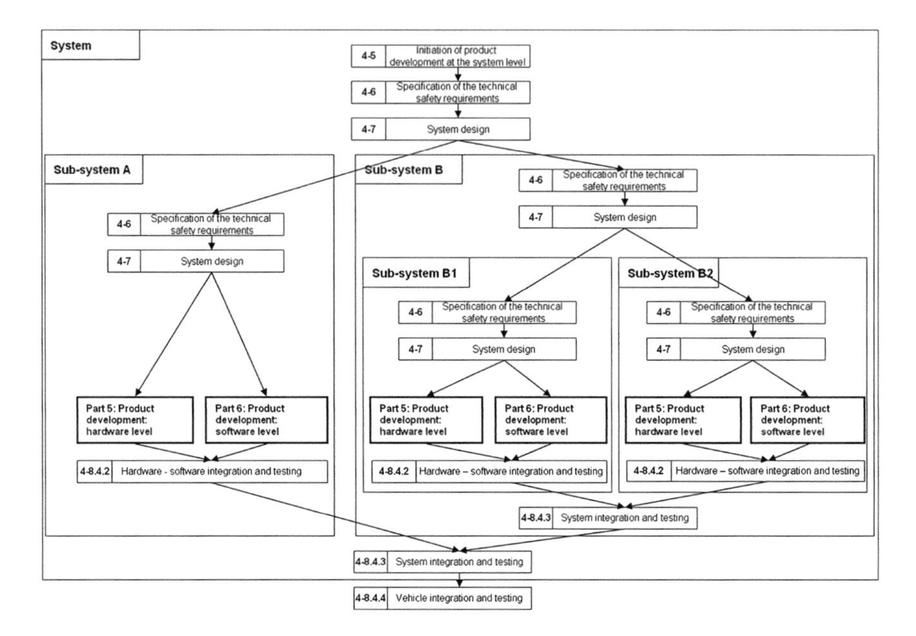

Figure from ISO 26262: Example of a product development at the system level (Source: ISO 26262, Part 4, Fig. 3)

The Fig. 3 shows, that integrations are depending on the defined horizontal layer, but also within a horizontal layer elements like different software components should be integrated in a multistep approach. Method like continuous integration need at least a higher degree of planning activities and tooling to control the integration steps becomes essential.

The most common verification method is testing. Test methods have different aims and are hence grouped differently. Consequently there are tests, which support the development of requirements. The tests during the development of requirements derive mainly from analysis (like FMEAs) or other verifications.

ISO 26262, *Part 4*, *Clause 7*:

7.4.3.7 This requirement applies to ASILs (A), (B), (C), and (D), in accordance with 4.3: In order to avoid failures resulting from high complexity, the architectural design shall exhibit the following properties by use of the principles in Table 2:

a) modularity; and
b) adequate level of granularity; and
c) simplicity

Table 2—Properties of modular system design

Properties		ASIL			
		A	B	C	D
1	Hierarchical design	+	+	++	++
2	Precisely defined interfaces	+	+	+	+
3	Avoidance of unnecessary complexity of hardware components and software components	+	+	+	+
4	Avoidance of unnecessary complexity of interfaces	+	+	+	+
5	Maintainability during service	+	+	+	+
6	Testability during development and operation	+	+	++	++

Figure from ISO 26262: Table 2—Properties of modular system design (Source: ISO 26262, Part 4, Table 2)

In Table 2 line 6 "Testability during the development and operations" for ASIL C and D continuously testability is required. During operation of the product built-in self-tests are considered, but also during the operation of prototyping and sample testing.

Surprisingly testability during development and bevor sample delivery to OEMS or higher TIERs are normally required by the given quality standards. The table should be interpreted, that for ASIL C and D applications a proper hierarchical and modular design is compulsory and adequate integration with adequate tests of horizontal and vertical interfaces are strongly required.

For the verification of the system design ISO 26262 requires the following:

ISO 26262, Part 4, Clause 7.4.8.1:

7.4.8.1 The system design shall be verified for compliance and completeness with regard to the technical safety concept using the verification methods listed in Table 3

***Figure**: ISO 26262, Part 4, Table 3—System design verification*

Methods		ASIL			
		A	B	C	D
1a	System design inspection[a]	+	++	++	++
1b	System design walkthrough[a]	++	+	o	o
2a	Simulation[b]	+	+	++	++
2b	System prototyping and vehicle tests[b]	+	+	++	++
3	System design analyses[c]	see Table 1			

a Methods 1a and 1b serve as check of complete and correct implementation of the technical safety requirements.

b Methods 2a and 2b can be used advantageously as a fault injection technique.

c For conducting safety analyses, see ISO 26262-9: —, Clause 8 (Safety analyses).

Figure from ISO 26262: *Table 3*—System design verification (Source: ISO 26262, Part 4, Table 3)

> *NOTE Anomalies and incompleteness identified between the system design, regarding the technical safety concept, will be reported in accordance with ISO 26262–2:—, Clause 5.4.2 (Safety culture)*

The note refers to part 2, here verifications are a mayor input for the "Confirmation Reviews" which is important input for the confirmation of the functional safety of the entire "ITEM".

The tables require many activities which are already required by quality standards like APQP, SPICE etc. consequently ISO 26262 expects the activities to be done, but not necessarily with the stringency of the context of a safety standard like ISO 26262. Especially

Similarly to the system we can find "Methods for the verification of software architecture design" in part 6, Table 6 in which requirements are stated similar to part 4 and additionally the control and data flow analyses. Table 3 (hardware verification) in part 5 shows the analogue requirements for the electronic hardware. This example shows that the analogy, which confirms, that ISO 26262 also during component development like software and hardware development considers a system development process.

In the Chap. 7 we will see that the entire verification needs to be planned so that under these requirements we can see a certain multiplications in the norm.

There are a lot of tables for the integration tests, which should support the test planning phase in the context of the integration into the respective horizontal levels.

The following tables are mentioned in part 4:

- Table 4—Methods for the test case development for integration tests
- Table 5—Correct implementation of technical safety requirement at the hardware-software level
- Table 6—Correct functional performance, accuracy and time behavior of safety mechanisms at the hardware-software level
- Table 7—Consistent and correct implementation of internal and external intersections at the hardware-software level
- Table 8—Efficiency of the diagnostic coverage of safety mechanisms at the hardware-software level
- Table 9—Level of robustness at the hardware-software level

Tables 10–14 show the methods for the requirements at the system level and Tables 15–19 at the vehicle level. Referring to the respective horizontal levels, the requirements and methods are supported by examples and references and differ from one another in detail.

For part 5 'Hardware' (Tables 10–12) and part 6 'Software' (Tables 9–16) there are also comparable tables. These also refer to the special features for the software and hardware development besides the adjustment of the respective horizontal levels. For their software a data and control flow analysis (for ASIL C and D required) is recommended in the tables. These analyses are not necessarily methods referring to the typical verification aims (complete, correct and consistent) they are rather comparable to the parallel to the architecture development required safety analysis.

Generally the methods in the tables can be grouped as follows:

- Methods for test case development
- Methods for tests to confirm correct implementation of the respective requirements
- Methods for tests of the performance, tolerances and timing behavior
- Methods for tests of the internal and external interfaces
- Methods for tests of the effectiveness of quality measures (e.g. assurance of design characteristics) and error control mechanism (e.g. safety mechanism)
- Methods for the robustness tests
- Methods for element specific analyses and tests

Tests are often distinguished between elements (components-, modules-, units, etc.) or integration tests. Most of the methods even the name give relevant information.

Typical element tests do questioning the input and output relation, the behavior in different environmental conditions or in case of different configurations.

Integration tests are always related to the interaction of the elements to be integrated in their specified environment and operating or application condition.

The verification will be iteratively called in the development cycle of ISO 26262 and change only in the level of abstraction in its scope, but the basic activities and the principles of methods remain similar. ISO 26262 describe the process iterations and the application in the different level of abstraction as follow:

ISO 26262, Part 8, Clause 9:

9.2 General

9.2.1 Verification is applicable to the following phases of the safety lifecycle

- *the concept phase, here verification ensures that the concept is correct, complete and consistent with respect to the boundary conditions of the item, and that the defined boundary conditions themselves are correct, complete and consistent, so that the concept can be realised.*
- *the product development phase, here verification is conducted in different forms*:
- *in the design phases, verification is the evaluation of the work products, such as requirement specification, architectural design, models, or software code, thus ensuring that they comply with previously established requirements for correctness, completeness and consistency. Evaluation can be performed by review, simulation or analysis techniques. The evaluation is planned, specified, executed and documented in a systematic manner.*

NOTE 1 Design phases are ISO 26262–4, Clause 7 (System design), ISO 26262–5, Clause 7 (Hardware design), ISO 26262–6, Clause 7 (Software architectural design) and ISO 26262–6, Clause 8 (Software unit design and implementation)

- *in the test phases, verification is the evaluation of the work products within a test environment to ensure that they comply with their requirements. The tests are planned, specified, executed, evaluated and documented in a systematic manner.*
- *the production and operation phase, here verification ensures that:*
- *the safety requirements are appropriately realised in the production process, user manuals and repair and maintenance instructions; and*
- *the safety-related properties of the item are met by the application of control measures within the production process.*

NOTE 2 This is a generic verification process that is instantiated by phases of the safety lifecycle in ISO 26262-3:—, ISO 262624:—, ISO 26262–5:—, ISO 26262–6:— and ISO 26262–7:—. Safety validation is not addressed by this process. See ISO 26262–4:—, Clause 9 (Safety validation), for further details

A systematic integration can only lead to success if already verified elements are integrated. Since this cannot always be the case, the integration will always happen iteratively when the verification results are available. Therefore, the recursion strategy for the tests should not only be referred to the components or element tests but the integration needs to be at least planned in the next higher level (Figure 6.1).

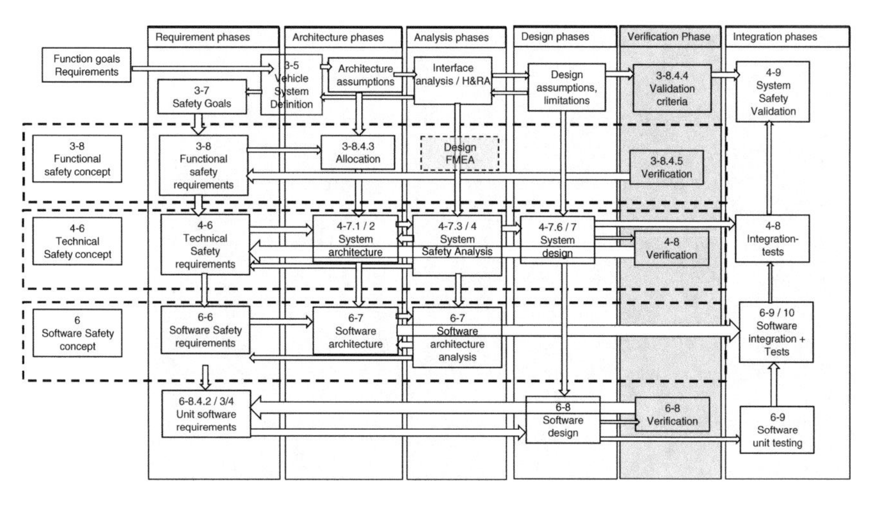

Fig. 6.1 Verifications between design and integration phase

6.1.1 Basic Principles for Verifications and Tests

Any verification shall be planned. ISO 26262 requires at least verifications after any phase of the development cycle. The information flow in the figure above shows how it fits into a typical hierarchical design.

ISO 26262, Part 8, Clause 9:

9.4.1 Verification planning

9.4.2.1 The verification planning shall be carried out for each phase and subphase of the safety lifecycle and shall address the following

a) the content of the work products to be verified
b) the methods used for verification,

NOTE 1 Methods for verification include review, walk-through, inspection, model-checking, simulation, engineering analyses, demonstration, and testing. Typically verification applies a combination of these and other methods

c) the pass and fail criteria for the verification,
d) the verification environment, if applicable,

EXAMPLE A verification environment can be a test or simulation environment

e) the tools used for verification, if applicable,
f) the actions to be taken if anomalies are detected, and
g the regression strategy.

NOTE 3 A regression strategy specifies how verification is repeated after changes have been made to the item or element. Verification can be repeated fully or partially and can include other items or elements that might affect the results of the verification

9.4.1.2 The planning of verification should consider the following

a) the adequacy of the verification methods to be applied,
b) the complexity of the work product to be verified,
c) prior experiences related to the verification of the subject material, and

NOTE This includes service history as well as the degree to which a proven in use argument has been achieved

d) the degree of maturity of the technologies used, or the risks associated with the use of these technologies.

After planning of the verification the activities of the verification shall be specified. There 3 methods considered, but only testing is detailed. ISO 26262 defines the following requirements:

ISO 26262, Part 8, Clause 9:

9.4.2 Verification specification

9.4.2.1 The verification specification shall select and specify the methods to be used for the verification, and shall include

a) review or analysis checklists; or
b) simulation scenarios; or
c) test cases, test data and test objects.

9.4.2.2 For testing, the specification of each test case shall include the following

a) a unique identification,
b) the reference to the version of the associated work product to be verified,
c) the preconditions and configurations,

NOTE 1 If a complete verification of the possible configurations of a work product (e.g. variants of a system) is not feasible, a reasonable subset is selected (e.g. minimum or maximum functionality configurations of a system)

d) the environmental conditions, if appropriate,

NOTE 2 Environmental conditions relate to the physical properties (e.g. temperature) of the surroundings in which the test is conducted or is simulated as part of the test

e) the input data, their time sequence and their values, and
f) the expected behaviour which includes output data, acceptable ranges of output values, time behaviour and tolerance behaviour

NOTE 3 When specifying the expected behaviour, it might be necessary to specify the initial output data in order to detect changes

NOTE 4 To avoid the redundant specification and storage of preconditions, configurations and environmental conditions used for various test cases, the use of an unambiguous reference to such data is recommended

The requirements show, that basis for most test activities based on verifications. About reviewing and check-lists no further details are mentioned. Even about simulation scenarios no further requirements are defined. At least for verifications based on simulations similar requirements as for testing should be considered. The intent of ISO 26262 is not to become another test standard, but during the development of test scenario and test cases a lot of methodology could be considered from other standards.

Also ISO 26262 sees 3 methods of grouping, but it is more considered that these groups should be applied within the same level of abstraction and depending on organizational interfaces. ISO 26262 defines the following requirements:

ISO 26262, Part 8, Clause 9:

9.4.2.3 For testing, test cases shall be grouped according to the test methods to be applied. For each test method, in addition to the test cases, the following shall be specified:

a) the test environment,
b) the logical and temporal dependencies, and
c) the resources.

All verifications have to be performed as planned but also with a dedicated result and during the execution of tests a target oriented approach is defined. After verification a certain evaluation (which is more or less a verification of the verification) is also required. ISO 26262 defines the following requirements:

ISO 26262, Part 8, Clause 9:

9.4.3 Verification execution and evaluation
9.4.3.1 The verification shall be executed as planned in accordance with 9.4.1 and specified in accordance with 9.4.2
9.4.3.2 The evaluation of the verification results shall contain the following information

d) the unique identification of the verified work product,
e) the reference to the verification plan and verification specification,
f) the configuration of the verification environment and verification tools used, and the calibration data used during the evaluation, if applicable,
g) the level of compliance of the verification results with the expected results,
h) an unambiguous statement of whether the verification passed or failed; if the verification failed the statement shall include the rationale for failure and suggestions for changes in the verified work product, and

NOTE The verification is evaluated according to the criteria for completion and termination of the verification (see 9.4.1.1 c) and to the expected verification results

i) the reasons for any verification steps not executed.

All tool settings and also the verification environment have to be recorded as part of the verification measures. These requirements are not only valid for testing, especially if simulations are applied for verifications, the records of this information are very important to trace the result.

Even more important is the trace for the requirement (g), the level of compliance. This requirement is a hidden requirement for the confirmation review in Part 2 of ISO 26262. Questioned could even be, if the assessment about the degree of compliant is not more a topic for a "Functional Safety Assessment". The last requirement (h) leads to the result and the consequences of the entire verification activity. The decision must be taken if by the defined change management process iteration should be considered or if the result is sufficient to accept it, which is at the end also a typical assessment topic.

These requirements from ISO 26262 address planning, execution and assessment of the verification. It facilitates the verification if these requirements are already considered for the entire process of verification and testing as the mayor method for the process step of verification. If ISO 26262 is already considered as a process, verifications are very often required, especially in case of distributed development and in case of many system levels. Questioned could be, if by defining a separate verification process, the efforts could be reduced and synergies found.

6.1.2 Verification based on Safety Analyses

Safety analyses are principally only special methods for the verification. Especially the different FMEA methods support the verification of systems, components or any other element type.

A System-FMEA primarily supports the verification of the architecture, functions and indirect also requirements and their allocations on functions as well as logical or technical elements. A Design-FMEA questions the correct design or also the realization of elements. In this case we often start with design drafts and in the later iterations the developments will be integrated more and more and the element characteristics maturity increases with any iteration. Therefore, the Design-FMEA primarily supports the design verification and is completed usually by a design review (especially the so-called Toyota FMEA, DRBFM—Design Review Based on Failure Modes focuses on design reviews by experts). A Process-FMEA generally analyzes the production process. However, technically it would be possible to analyze any random process with this method. We can find indications for that in Chap. 7 of this book (Process analysis for functional safety).

Each FMEA standard includes the additional requirement that the result of an FMEA needs to be checked again in regards to the target of the analysis. A final review of the FMEA is formally part of each FMEA method.

The following verifications can be supported by safety analyses:

Completeness of the relevant safety goals

Safety goals are primarily formulated as follows: "Avoid that a possible malfunction violates a safety goal". All possible malfunctions could be systematically analyzed if they have the potential to violate safety goals. All functions on the vehicle system level of the ITEM could be analyzed for potential malfunctions. The malfunction could be considered as failure or errors. The negation of the safety goals could be considered as top-failure of the FMEA so that they could be handled as possible failure effects. If the technical errors from the Technical Safety Concept are considered as failure causes, and malfunctions, errors or failure in the Functional Safety Concept as failure type a typical 3 level FMEA could be considered. The FMEA could demonstrate completeness for

- The Technical and Functional Safety Concept
- considered malfunctions with the potential to violate safety goals

and means a mayor input for the verification of Functional and Technical Safety Concept.

Completeness of functional specification of safety mechanism or any other safety-related function

This analysis is based on the function net of the VDA-FMEA. In animated SYSML-tools the function net of the VDA-FMEA could be even used more effectively. The relevant functions need to be decomposed and allocated to structure elements in the structure net of the VDA-FMEA. The structure net could be any decomposition of the architecture based on technical, functional or logical blocks. The functional decomposition should be performed only on a defined horizontal level of abstraction.

It is very similar to the function decomposition in a System-FMEA. It can be performed at components level for hard- and software, and on any system level. This analysis is even recommended within semiconductor structures. The basic principle of the analysis is the identification of signal chains, which has previously been described by Robert Lusser over 80 years ago. In addition, another method called "FAST, Functional Analysis System Technique" exists, which describes this derivation of functions at a lower element structure and their inductive analyzing approach.

Completeness of higher level requirements to lower level requirements based on functional decomposition.

For these analyses we consider the pure functions without any allocation to elements. Maybe this analysis could be performed before the previous described analysis, but if it could be done independently no systematic errors lead to common failure. An already verified function in a higher horizontal level of abstraction have to be completely derived to a lower level (e.g. from vehicle level to system level or from higher system level to lower or component levels). In a VDA-FMEA we could again test through the function net, whether the functions in the lower level (e.g. components in SW or HW) are completely displayed at the system level (if you could show all signal chains of the system level, completeness of the specification of Hardware-Software-Interface (HSI) could be analyzed). This analysis also supports the analysis of dependent failures, since the dependencies in lower levels are displayed into context with reference to the dependency in the upper levels. This cannot only be analyzed for functional dependencies; also physical dependencies (for example temperature influences, EMC) or energy dependencies can be analyzed that way. Signal chains are data-flows, but energy, temperature etc. are also physical flows, which could be verified by even physical laws such as law of conservation of energy or Kirchhoff's law for voltage or current etc. The function decomposition could be done top-down (in this case it is more a positive fault-tree analysis (FTA)) or from bottom-up and it could be considered more as an inductive analysis. Using SysML with the appropriate animation or test algorithms more detailed and automated analysis could be done with more detailed results in comparison to a VDA-FMEA. These algorithms can also be used for systematic function decomposition and its transparency modeled in the style of the method "SAFT, Structured Analysis Design Technique".

Consistency test of interfaces (Product decomposition)

The VDA-FMEA describes the interfaces for the considered product structure by the structure net. The function net represents the functional interfaces between the elements of the structure net. A VDA-FMEA could structure by different element types, so that the structure could base on a functional, logical or a technical decomposition. That leads to the consequence, that for all 3 structures different FMEAs become necessary, because these are 3 different analysis. Virtual interfaces such as the RTE (run-time environment) could be considered as logical interfaces, so that also errors of hypervisor, priority management and scheduling functions in software could be analyzed. By comparing of resulting of the functional structures we can analyze their dependencies and their consistency over the different architectural views also by distinguishing of different levels of horizontal abstractions. If interfaces or dependencies in the system differ from those on lower level e.g. the components level, the inconsistencies need to be solved. It would be recommended to use SysML-tools for such different views and level of abstractions, so that automated checker could analyze the inconsistency. By using typical FMEA templates or even pure FMEA-tools, the manual review becomes rather complex.

Completeness of the failure possibilities considered

Even for a deductive analysis completeness analysis all error impacts is an important argument. Of course, each possible error, fault or malfunction found will help to improve the quality of a product, but for safety, completeness is required. This is why the failure analysis in the VDA-FMEA is at step three after the product and function decomposition. This means that for each function of a structure element the possible malfunctions need to be identified. For verification of the safety-relevant requirements it is important to first analyze, whether the possible malfunction, which can lead to a malfunctional behavior, have been completely identified. For a mere functional analysis stating that data flow, signals or information just could have 2 error states:

- no function or
- incorrect function

Based on the fact on a basic level completeness of an analysis could be argued. In the more in-depth analyses we can analysis whether the following malfunctions functions have been considered for the completeness argument:

- no function
- unexpected function (crosstalk of other systems)
- systematically falsified function or information (i.e. signal drift)
- sporadically or unexpected incorrect function or information
- module or element has not been implemented, addressed or considered
- continuously operation as specified for functions or elements such as interruption free operation, no oscillation, intermediate errors, random or sporadic faults etc.
- incorrect time behavior

Those are also typical questions for deductive methods such as HAZOP or the fault tree analysis (FTA). The malfunctions (or error modes) also show in the tables of ISO 26262, part 5, Appendix D, which represent the foundation for the diagnostic coverage. Which of those error modes are relevant depends on the requirements and their context which are imposed on the functions. This is why at this in-depth level not only the architecture is analyzed but also the design and the realization. Therefore, such analyses are often on lower component level and performed by means of a Design-FMEA and define the basis for the design verification and validation (DV).

Completeness of the considered single-point faults

This is the classical domain of the FMEA. At this point all possible malfunctions of a respective level are evaluated whether they can propagate to a given safety goal or if they are a possible cause of a failure effect, that violates safety goals. Here a classical FMEA could claim completeness related to the considered scope of analysis.

Complete consideration of dual-point-faults

Multiple-point faults always build high permutations referring to its influence factors. This is why even for simple systems the multiple-point failure analysis is considerable a challenge. However, if safety mechanisms design as safety-barriers, which should prevent fault propagation and their penetration through barriers, the analysis of the individual barriers become single-point fault analysis. Consequently the entire safety concept has to be designed based on multi-level safety barriers. So that a fault in a safety-related system could be systematically hindered propagating in horizontal and vertical direction. Consequently for ASIL C and D systems at least dual-barrier safety architecture becomes compulsory. Any possible fault need to break at least 2 safety-barriers in a safety-related system before they have the potential to violate considered safety goals. Consequently, it is more an architectural design approach to develop adequate safety architecture with such safety-barriers than a matter of analysis or verifications. The analysis should identify the gaps and the verification should show completeness, correctness and consistency of the safety-architecture and providing the confidence of the effectiveness of the applied safety measures.

Correctness of the safety goal

Safety goals could only verified against the intended function, for the intended use and possible malfunction in the intended environment. In order to verify or assess the correctness of safety goals the inputs have to be verified. Based on incomplete, incorrect or even inconsistent input not resulting activity could be lead to proper correct output. By performing the Hazard & Risk-Analysis also any possible hazard and the relation to driving situations and possible operating modes and possible transients have to be evaluated. Considering the result of the Hazard & Risk Analysis and the assessed safety goal and its ASIL based on complete input as a deductive approach, an inductive verification could lead to a correctness statement. In the arguments before the possible malfunctions of were not used for verification. A typical event tree analysis (ETA) could show the relation between malfunctions, functions and relevant driving, systems or operating modes in the intended environment, so that the correctness of safety goals could be evaluated by the answer to the question, if all possible malfunctions are sufficiently controlled in case of fulfilling the safety goals.

6.1.3 Verification of Diverse Objectives such as Safety and Security

Design, architecture of a product or even the requirements should be consistent for any characteristics or features during different steps of the development cycle. Completeness and correctness are in general basics to assess consistency. If the design, the architecture or requirements are incorrect or incomplete, any corrected or added characteristics could violate the consistency requirements.

Questionable is, if completeness or correctness verification could be done for could be done in one step for diverse objective like safety and security. Consequently both have to be complete and correct before their common consistency could be verified. If safety and security mechanism block each other the implementations of the mechanism make whether neither safe nor secure; maybe with some compromises but that violates very often objectives like performance, availability or other characteristics.

The following table 6.1 could be an example for a joined concept development process.

Table 6.1 Verification of Safety and Security

Process step	Activities for Safety	Activities for Security	Common Activity
tool qualification	according to ISO 26262	analysis security impacts	agreement of measures
Security in development environment	respect and planning of safety activities based on the security constraints for development and production	development of a security concept for development and production	check the adequacy of the measures
Item Definition	analysis of system boundaries	authorization concept	
Hazard Analysis & Risk Assessment	risk identification and ASIL	threats identification	
Functional Safety Concept	safety measures and allocation	measures definition	
Verification of FSC	verify of measures	verify of measures	evaluating the ability of coexistence of both sets of measures
System specification / Architecture	define requirements, architecture, behaviour and internal interfaces	requirements, implementation concept	allocating security mechanisms on product architecture
analysis	inductive / deductive	effectiveness analysis	Analysis - error behaviour of all functions and mechanisms
System - Design	define system, parameters, requirements for components	derivation and allocation of security mechanisms	Verify the coexistence capability of both set of measures
Verification of components requirements	verify feasibility, completeness, correctness, consistency of standards for component	verify feasibility, completeness, correctness, consistency of requirements for components	analyse consistency of both sets of requirements.
Component specification / Architecture	define requirements, architecture, behaviour and internal interfaces	requirements, implementation concept	allocating of security mechanisms to component architecture
Analysis of component	inductive / deductive	effectiveness analysis	analysis - error behaviour of all functions and mechanisms
Components - Design	component design, parameters, requirements for components	derivation and allocation of security mechanisms	verify the coexistence capability of both set of measures
Verification of the components prior to implementation / deployment / realisation	verify feasibility, completeness, correctness, consistency of requirements before implementation	verify feasibility, completeness, correctness, consistency of requirements for implementation	analyse consistency of both sets of requirements.
implementation / deployment / realisation	implement as specified	implement as specified	verify the specification-compliant implementation
Integration / Test	tests according to specification	tests according to specification	tests of ability of coexistence and effectiveness of both sets of mechanisms.

6.1.4 Test Methods

The objective of part 4, chapter 8, 'Integration and Test' have previously been discussed in the planning of the architecture. 3 Integration phases are considered in ISO 26262.

ISO 26262, Part 4, Clause 8:

8.1 Objectives

8.1.1 The integration and testing phase comprises three phases and two primary goals as described below: The first phase is the integration of the hardware and software of each element that the item comprises. The second phase is the integration of the elements that comprise an item to form a complete system. The third phase is the integration of the item with other systems within a vehicle and with the vehicle itself

8.1.2 The first objective of the integration process is to test compliance with each safety requirement in accordance with its specification and ASIL classification

8.1.3 The second objective is to verify that the "System design" covering the safety requirements (see Clause 7 (System design)) are correctly implemented by the entire item

As a consequence, three horizontal system levels have to be developed, in which elements are integrated hierarchically up until the vehicle integration. The following two objectives are directed to methods how to integrate in the dedicated horizontal level of abstraction.

Requirements based testing is the first objective for any level of integration. The tests should show the correct implementation of the given requirements related to the product but also relevant by the given standards, particularly here ISO 26262.

Since a system according to ISO 26262 needs to be a hierarchically structured, the following two basic test types have to be considered:

- element test
- integration or interface tests

In the acceding branch of the V-cycle the objective is not anymore the verification of requirements, now the realized product have to be verified. Here are not anymore architectural interfaces relevant for the integration; here the real product interfaces are relevant. Of course all architectural interfaces shall be still transparent to the elements of the realized product, but in some cases not all interfaces could be animated to verify the entire specification space. So only a repetition of the tests during requirement development is not sufficient.

ISO 26262 requires in part 4 on system level, but the principles of integration are also applicable for element integration within components or even during development of semiconductors even inside such hybrid devices.

ISO 26262 recommends or requires the requirement development or at the least for its verification, to test the planning or testability of the correct implementation through the realization of the requirements. This means that if a requirement is developed, a concept is needed beforehand showing how the correct implementation of the requirement can be proven for the product developed. If the test planning had started at a later time in the development, we could probably systematically cause a change in the developed requirements. Key words such as "Design2Test", DoE (Design of Experiment), requirement based test management or risk based testing describe possible test methods.

6.1.5 Integration of Technical Elements

The different levels of the integration are used for the verification of the interfaces of the relevant elements. The typical verification criteria for the interfaces build the foundation for the applied methodology. The objective of such tests are whether the interfaces have been developed completely, consistently, correct and if sufficiently transparence is given for the safety case.

In real systems or products systems do not consist just of 1 software and 1 hardware component. Elements of other technology are mixed with electronic, such as connectors, printed-circuit-board, harness etc. Also inside the microcontroller the hardware-software-interface exists not in single blocks. Nearly any function in software in any control cycle passes interfaces of different technology by any software instruction. The challenge is to find a proper and traceable integration strategy, so that the intended functions and their performance could be confirmed and the safety verification leads to positive results. If the tension between performance, availability and safety leads to endless process iterations due to negative results of the verification, real safety cases could not be assessed. In case of fail-operational systems during any integration cycle the availability of the intended function in any integration level need to be verified, so that dependent failure do not lead failure of the intended function. Furthermore the integration strategy should respect also the given security requirements. If the entire non-functional requirements are not harmonized during integration any verification will fail.

Beside hardware, software and other technology also other elements need to be considered during integration. In ISO 26262 we can find 3 categories of elements, which impose different requirements to their integration.

Safety elements developed out of the context of specific item or vehicle system (SEooc, Safety Element out of Context)

In this category we can almost find all elements and components, which are integrated into a vehicle. Microcontroller, software components and even entire vehicle systems such as electronic brake systems are not developed for a certain vehicle with certain driving dynamics data, they are developed based on experiences or market analysis. Because of that, a lot of interfaces (even electric interfaces) are standardized and even those standard interfaces are not always derived from typical automotive applications. Consequently the architecture could not developed as a top-down approach, specific interfaces have to be considered as constraints. As more unique the interfaces could be design, as wider is their range of application. Microcontroller, communication systems such as CAN-Bus, connectors nearly any hardware will provide constraints for the system design. In order to find a modular design and a wide range of use for such SEooCs, assumptions of higher level requirements have to be developed so that those elements could later be successfully integrated into a specific vehicle. From the top-level, even safety goals and the possible ASIL have to be assumed, so that a hierarchical design could be assured to the lower levels down to the realization.

During the integration of the SEooC all assumption have to be mapped to given requirements and constraints of the target vehicle, so that the results of the SEooC development could become valid also for the safety case of the target item.

Qualified components

ISO 26262 addresses the qualification of software and hardware elements in part 8 respectively separate. Both elements have the same challenges. It have to be assumed that elements to be qualified have not been developed according to ISO 26262. Since ISO 26262 is still pretty new, there are not that many components, which are developed according to this norm. Therefore, before its publication of course other standards or norms have been used instead. If a component has been developed according to a different safety standard, we can generally assume that there is an certain level of safety documentation. However, whether the failure propagation of this component in a vehicle system is the same as in an airplane or in a stationary power plant and whether the time behavior is sufficient in combination with other safety mechanisms, can become challenging questions. Especially the required multiple-point failure control for ASIL C and higher becomes challenging. Since hardware components are physically describable, ISO 26262 also allows such a qualification for new components. ISO 26262 elements are not made of a specific substance or material. Any basic element need somehow qualified in order to be suitable for a safety-related application. Arguments for their trustful functions and adequate evidence need to be provided.

It is not the aim of the automobile industry to one day develop resistors according to ISO 26262. However, this is not the case for software elements; in this case the norm suggests that this type of qualification should not be used for new developments. Doesn't the software often behave different in a different microcontroller? Are core operations for the different code instructions so unique? Could compiler settings from one controller to another controller lead to the same safe functioning? Even Autosar could not assure a sufficient safe and consistent environment for a safety-related application software.

Proven elements, proven in use (PIU)

This is one of the most difficult topics of safety engineering. First of all, a lot of experienced developers say that there have not been any previous safety risks by using proven elements in vehicles. Why is a system now no longer "safe" only because such a norm has been published? The risk has already been described before with qualified elements; we do not know whether the case of application, the integration environment and the error propagation happening at the interfaces are actually that identical. According to ISO 26262, PIU is a "Black-Box-Approach", meaning we do not know the entire inner structure and behavior of the elements or the candidate to be reused. Safety should be assessed purely according to the characteristics of the outer boundary. The norm does not support this since it requires that the performance and the frequencies of errors need to be quantified based on field experiences. This quantification however should be based on a comparable case of application in a comparable environment. Since microcontrollers especially change constantly, this signifies an enormous challenge for software elements.

6.2 Safety Validation

Validation need to be considered different in the context of safety validation or any other validation activity. Furthermore safety validation in ISO 26262 is a specific activity and the discussed topic for verification to see validation as a methodology. Validation in general is often described as the confirmation of targets. A certain general validity is required and man could say: "Validation is the proof that targets are reached reproducibly".

ISO 26262 sees safety validation on vehicle level, but safety validation is not considered as a single activity. Safety validation should accompanied the whole safety process until the end of the development phase.

ISO 26262, Part 3, Clause 8:

8.4.4	*Validation criteria*
8.4.4.1	*The acceptance criteria for safety validation of the item shall be specified based on the functional safety requirements*
NOTE	*For further requirements on detailing the criteria and a list of characteristics to be validated see ISO 26262–4: 6.4.6.2 and ISO 26262–4: 9.4.3.2*

The note gives clear hind that the safety manager has to plan further intermediate validation activities during planning the activities (Part 4, Clause 6.4.6.2) for product development on system level. Of course a similar validation steps meaningful during component development, but those are not explicitly required by the standard. Those intermediate validations are typical characteristics of today's agile development approaches, but they also are inherent part of a spiral process or automotive typical APQP activities.

However, in contrast to verification, validation still has kind of a blurry character since targets are often not as precisely formulated as requirements, which are generally verified according to ISO 26262. In the automobile industry we can often find the following definition: "The customer requirements are validated, but the requirements for example in the product and technical specifications are often verified." Whereby the product specification is seen as the formulated wish of the customer and can once again be validated.

The following aspects are associated with the term validation:

- Latin: validus; strong, effective, healthy
- Validity: weight of a statement, investigation, theory or premise
- Validation is a method of communication with dementia patients
- Validation: Proof that a process, a system and/or the production of a active substance reproducibly fulfills the requirements in practical use
- Validation for a semiconductor says that it can be produced according to the specifications
- Validation: external proof of large-scale projects and theirs sustainability reports
- Validation in computer science is the proof that a system meets the requirements in practice
- Validator: A method or program, which should confirm the verification with respect to a standard
- Validation or testing of the validity of statistical values or their plausibility
- Method validation proves that an analytical method is suitable for its purpose

- Validation often describes a statistical proof
- Validation of educational attainment
- Model validation should show that the system developed through the implementation of a model reflects reality sufficiently precisely

The addressed interpretations above also show all aspects, which play a role for functional safety. However, through the multifaceted nature of the term it is difficult to find a definition of the general term 'validation'. Therefore, this term has a rather restricted meaning in ISO 26262. All other validation aspects are paraphrased with verification or analysis.

In part 4, Chap. 9 the safety validation activity is described as follows:

ISO 26262, Part 4, Clause 9:

9.1 Objectives

9.11 The first objective is to provide evidence of compliance with the safety goals and that the functional safety concepts are appropriate for the functional safety of the item

9.1.2 The second objective is to provide evidence that the safety goals are correct, complete and fully achieved at the vehicle level

Two objectives are defined for safety validation; the first is the evidence that the safety goals are considered adequately in the context of the functional safety concept and the defined item. The second objective asks for the evidence that the safety goals themselves are correct and achieved on vehicle level. The hope of any safety validation is, to proof that the vehicle is safe as such, but ISO 26262 could provide support on the evidence of functional safety for E/E-Systems. The safety-live-cycle in ISO 26262, part 2 shows, that external measures and also measures of other technology have to be considered during safety validation. In "9.2 General" the relation to other activities are detailed.

ISO 26262, Part 4, Clause 9.2:

9.2 General

9.2.1 The purpose of the preceding verification activities (e.g. design verification, safety analyses, hardware, software, and item integration and testing) is to provide evidence that the results of each particular activity comply with the specified requirements

9.2.2 The validation of the integrated item in representative vehicle(s) aims to provide evidence of appropriateness for the intended use and aims to confirm the adequacy of the safety measures for a class or set of vehicles. Safety validation does cover assurance, that the safety goals are sufficient and have been achieved, based on examination and tests

The clause 9.2 defines more specifically how ISO 26262 sees the relation to verification and integration measures.

For example, examining whether a goal has been achieved, this includes the validation of the safety goals in ISO 26262. A test a for example, if the SW fits into the microcontroller would be a verification after integration. Methodically you checked but against the requirements of the higher level through the intermediate step, whether they were consistent, correct and completely achieved. A second step would be whether the goal of a safe integration of hardware and software has been achieved.

Thus it is also in the second interpretation, the request itself is really what is required, according to ISO 26262 in deriving the functional safety concept for technical safety concept through to the component requirements. Here ISO 26262 calls the activity also verification of requirements.

The mayor goals for validation based on underlying verifications such as:

- The goals are achieved
- The demands were the right
- The requirements were implemented correctly and adequately.

If this leads to a consistency and completeness of reasoning, creating a sufficient transparency should be no problem. If the activities and the methods used extensively documented, so that an auditor could confirm the adequateness of the activities and an assessor should be able to confirm the achievement of functional safety.

This means that all verifications are receipt confirmations that all relevant requirements and specifications are correctly implemented for a certain system on vehicle level for a specific vehicle or vehicle class. The safety validation should provide the final argument for functional safety on vehicle level.

6.3 Model Based Development

Model based developed is a heavily discussed topic in automotive industry. Most frequently this term means the automated code generation. However, simulations and automated checker functions instead of manual reviews are used more and more in context of functional safety especially for verification activities. Since for the verification we need to state the subject of the verification in an abstracted form, it is recommended, to let a model develop and mature alongside the product development. Whether the aim is to really create a complete model, which reflects the entire product in its integration environment, or the models should be adjusted for their purpose or even developed independently, depends on various factors. It is important however, that the models, which are used in development, are also considered in the context of project planning. The question is: "Which parts of the necessary activities can be automated and what purpose does the model serve?".

At the vehicle level (compare information flow with figures in previous chapters) a model would be highly useful since all assumptions can already be validated at the model in the design phase.

In the initial phase of the development it is very clear that models are used in order to better understand requirements or dynamic behavior can be described in the first process iterations. On the descending branch of the V-model, models are often abstractions of the corresponding product before its realization with which it can be proven whether the products can be integrated sufficiently in the ascending branch of the V-model. These models focus on describing the development in a way that the behavior corresponds with the product developed in an abstracted environment. Besides the failure analysis these models are often used as a basis for test benches (e.g. HIL, Hardware in the Loop), in order to develop automated tests. The realization model is often not derived from the requirement model but developed independently. The benefit is that the tester does not need to be involved in the requirement development and thus unbiased testing can be ensured. However, speaking on the verification or the validation of the models or their consistency, we can see that the significance of such models is limited. Furthermore, the verification of the requirements will be difficult referring to the correct implementation for inconsistent models. The benefits of such independently developed and validated models are the insights at the consistency check. At this point of course the respective systematic failure will be apparent. A parallel development and maturing of the model and the systematic verification or validation of the model against requirements as well as the previously developed and determined characteristics of the final product would be advisable. This of course can only apply for reduced abstractions. A complete modeling of electronic or even of the microcontroller used is possible nowadays but still very complex (Fig. 6.2 and 6.3).

The model has to achieve the intended target maturity until the development of individual components. This means that the model should comply with the relevant requirements by 100 %, which means complete. The integration of the components, which are planned according to the model and which correctness can be argued with the help of models, reaches its complete maturity with the finale validation. These requirements however can only apply if no change in the requirements is allowed during the development. However, based on the architecture, influences of changes can be explained. The model is going to support the influence analysis, especially for the technical behavior and the dynamic effects.

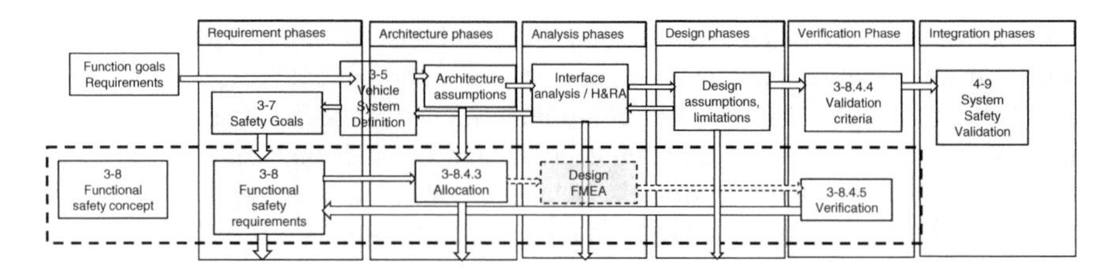

Fig. 6.2 Information flow on vehicle level engineering

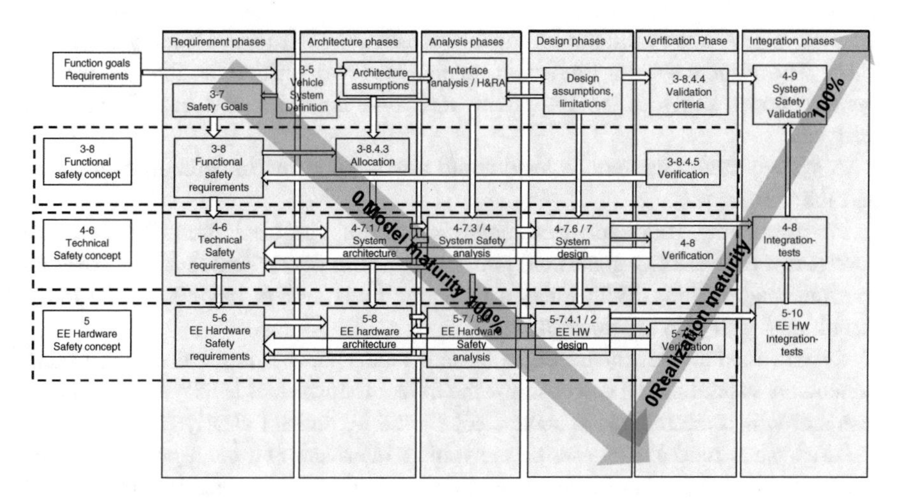

Fig. 6.3 Model maturity compared to maturity of the product developed

6.3.1 Models for Functional Safety

Vehicle model: Such a model can show the behavior of a vehicle in the context of driving situations. It can also illustrate the vehicle reaction based on possible malfunctions of systems on vehicle level, which needs to be integrated. Such models support the analysis and verification of the item including considerations of boundary condition, limitations as well as the requirement analysis for the intended functions. Especially if the intended function itself becomes safety-related, the necessary analysis to verify them within its intended range of use and environment could be performed, before any function is integrated in a vehicle. Such a model can support the hazard and risk analysis and also provide essential information for their verification. For the model it doesn't matter if the effect of malfunctions or mere functional effects are verified. This model could provide arguments for completeness for the verification of the functional safety concept. The derivation of functional safety requirements from the safety goals against the safety architecture at this horizontal vehicle level and thus the allocation can be verified. On a functional level it is possible to simulate how certain functional safety mechanisms react on possible malfunctions of the system within the vehicle environment. Through respective timely simulations it is also possible to show the intensity of malfunctions during different failure tolerance time intervals, so that the failure tolerance time intervals can be analyzed and defined. These models can generally also support and provide arguments for the integration of the vehicle system and its verification and validation.

System models on vehicle level: This model would illustrate the behavior of the considered system on vehicle level but cannot uniquely show the behavior and the effects of the vehicle system or the vehicle reaction in the traffic environment. This is why the model would be suitable to verify the relevant malfunctions of the hazard and risk analysis. It would not be capable of verifying or even validating the safety goals. The vehicle system limits can be analyzed and verified, so that the vehicle system model provides an important verifiable input for the hazard and risk analysis.

A system model on vehicle level could also support the integration of the item and the verification.

For the safety validation this model could only be used limitedly since the correctness of the safety goals could be questioned in the context that they cover all relevant malfunctions. With today's modeling tools system models can be transferred very well into a vehicle model.

System model: The focuses of a system model are the interfaces to the components. A system model can describe the different horizontal levels, this is why the levels at which the models is abstracted, should be defined clearly (Fig. 6.4).

The system model can describe the vehicle interfaces and the aspects described under the vehicle system models would be valid. The components interfaces could be defined the same way so that the behavior of the components and their functions or functionalities can be described, analyzed, or verified. Furthermore, a model of the microcontroller could be used in order to describe, analyze or verify the hardware–software-interface and the behavior of the software within the microcontroller (environment). Nowadays even in the silicon models are widely used to describe the functionalities of semiconductors and validate those against the different realized samples. This means that a system approach is used in order to display the inner behavior of the silicon. ADL (architecture description languages) are widely used for that purpose.

Specifications, analyses and their verification and validation (validation as test against customer requirements) can be reasoned on the model level. For the failure

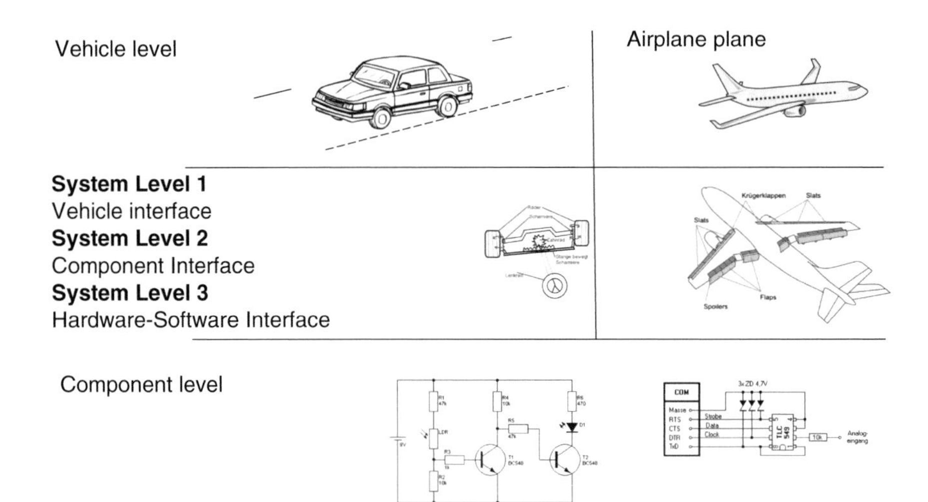

Fig. 6.4 Horizontal system level and type of interfaces

analysis of the model the same approach can be used as for the other hardware systems. The level of the desired functionality and its failures should be seen as the error type level. The causative level should be the level at which the measurable or observable anomalies of developed samples and typical systematic failures are describable. In this context it is primarily important that the model and the development mature continuously so that the model can also be respectively validated with each characteristic relevant for the product to be developed. Principally according to ISO 26262 the model validation is also a verification but in this case we will do without this for the semiconductor industry untypical term. A good model is sufficiently valid (suitable) and reproducible as a reference for the technical description in order to reason the analyses and verifications of the model.

Basically all models in safety technology are "System models". The entire ISO 26262 is based on a structure in which software and hardware components are also described through a systemic approach. Therefore, a combination of system elements is chosen that facilitates the implementation of the intended functionality.

Electronic Models: Electronic modeling is a fairly old discipline. Up until today SPICE (Simulation Program with Integrated Circuit Emphasis) is used as a foundation. SPICE (PSPICE is the PC version of SPICE) was developed in 1973 originally at the Electrical Engineering and Computer Sciences (EECS) department of the University of California in Berkeley. A comparable and even older algorithm is CANCER (Computer Analysis of Nonlinear Circuits Excluding Radiation). These algorithms were continuously improved and are even used nowadays as a basis for the description of electronic including semiconductors. Known system-modeling tools have integrated the SPICE algorithms. The term SPICE has nothing to do with the process assessment method, at which for examples Automotive-SPICE is based on today. It is only an example that electricians and software specialist do not have a systematic communication. Such SPICE algorithms can basically be embedded in each system environment so that the system and software intersection can also be described. SPICE algorithms can simulate the temperature, voltage and electricity behavior as well as mechanical influences on the behavior of electric components to each other. The models become especially significant since there are corresponding model libraries for all electric components, which also show the behavior of components in their integration environment.

Therefore, even antenna effects can be simulated through EMC malfunctions or drifts on transistors, which they themselves are not measurable with oscilloscopes. For EMC in our day's Maxwell equations are integrated in the tools. Furthermore, also heat behavior and its propagation within components and controlling units can be simulated. Especially for the analysis of dependent failure such a simulation can provide useful results. Since the error propagation can be simulated based on different effects, it is possible to detect failure cascades. In this case there is the possibility to reduce the causes of failure cascades or the propagation of failure cascades with adequate measures. By applying Kolmogorov-Smirnov or applying of algorithm based on those test method tests all kind of dependability could be simulated and analyzed. This means that the basic principles of system engineering including the error propagation are also applicable for these electronic models at the electronic and semiconductor level. If the reliability method, the principles of statistical failure distributions as well as the environment or integration profile are included in modeling (for example the Arrhenius approach), even quantitative safety analyses or importance analyses (Cut-Set analyses) can be illustrated in the model.

6.3.2 Foundation for Models

The foundation for models actually goes back to the questions of Parmenides (Greek philosopher), who pointed out 2500 years ago that not everything can be explained the way how it is observed. The moment influence parameters are included or left out, the observed behavior can change. Therefore, it can be shown that the form of abstraction of the model can be an essential basis for the significance referring to the development or the reality of the model (Fig. 6.5).

The P-diagram was previously discussed in the above-illustrated form in the 1950s. The diagram is based on the idea of energy transfer. The input value is seen as the 100 % ideal function. If 100 % of the input value could be transferred to 100 % of the output value, it would be an ideal system.

This does not exist in reality. The principle of thermodynamic says that a 100 % transformation is not possible; there is no such thing as a "Perpetuum mobile". This means we check which influences from the environment at the closed system cause which discrepancies at the output. Through this 100 % rule the reference to the requirements (has the model behavior been specified 100 %) as well as the development (is the observable behavior of the development 100 % explainable in the model) can be reasoned, so that it is possible to make statements referring to the level of abstraction and the model maturity. This was for example described in the Ford-FMEA handbook including the evaluation of interferences (robustness matrix). This P-diagram is used as basis for all meta-models, which can also describe the failure behavior of systems. This means that all descriptions and also behavior models show such a structure or a comparable one. Through the 100 % comparison we can reason completeness, so that an essential aim of the verification can be met. Those parameters of the P-diagram however need to consistently go through the entire model. If the model did not result from comparable (consistent) met models, consistency and completeness statements cannot be derived from the

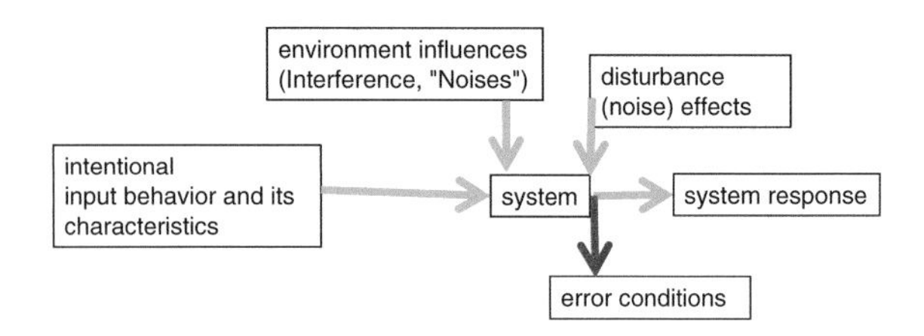

Fig. 6.5 P-Diagram (Parameter-Diagram) and its typical Parameter

model. Since both verification aims are the basis for safety technical correctness, also a safety technical correctness cannot be derived from the model without a consistent meta-model. All parameters of the product referring to requirements, design characteristics, architecture and the development itself, including all measures for the product as well as secondary the verification and validation, need to relate to the P-diagram and be technically consistently saved and archived. Otherwise change-management, configuration management and module management or baselining is only partially possible for safety technical systems. If each product, no matter at which horizontal abstraction level, is described through such P-diagrams, also the systematic approach is consistent in each horizontal level so that the system engineering principles can also be applied on software and hardware levels.

6.3.3 Model Based Safety Analysis

Since classic deductive and inductive analyses procedures such as FMEA or fault tree analyses (FTA) can only limitedly reflect the requirements or the development, model based safety analyses are important methods for the fulfillment of ISO 26262 requirements.

Of course it is only possible to analyze automatically, which has been automated or implemented into the model. Just like the idea of P-diagrams show, analyses and verifications are only as good as their foundations, which are available for the analysis or verification. The benefit of model-based analyses is that the processes could be automated and results can be supplied automated to further analyses and becomes repeatable.

An essential benefit of the model-based analysis is that with today's computers it is also possible to formalize illustrate the dynamic behavior as well as its safety technically analyzes. Especially the behavior in the case of failure or errors as well as in the transition from a static condition to different other states, which happens in different driving, systems or operation modes, needs to be considered. Failure behavior at such transitions of system conditions often lead to dangerous effects today's highly dynamic systems, which for example can no longer be controlled by the driver. Even effects, which at the transitions of system modes in the individual driving situations can cause danger, cannot be completely and systematically described or analyzed with the classic analysis methods. For a valid model we can configure the condition transitions with different parameters and model automated through a whole modeling range at each horizontal abstraction level (i.e. in the microcontroller, at the components level or at dedicated system levels). The observations of the output conditions and variations such as the failure reaction, as described in the P-diagram, allow systematic evaluations.

Therefore, failure combinations and even combinations of dynamic and static failures can be displayed. Thus for example a parameter principle of different drifts of a capacitor at the input of a transistor can demonstrate changed switching characteristics regarding different parameter fields. This example for a failure cascade can show a dual-point fault or even a single-point faults, which without such a simulation could only be illustrated with difficult and tedious tests and calculations. This shows that the simulation offers far more transparency for the analysis of dependent failure as well as multiple fault analysis. This example from electronics is of course also applicable for mechanics or software components as well as the system level. At the system level such failure combinations are hard to describe in connection with EMC influences. In this context a simulation can provide essential support. Whether the classic safety analysis method is now applied to the requirement model, the development model or the development itself should depend on the verifications and validations strategy.

However, the model-based safety analysis should first be seen as addition for the classic analysis methods. It would be worth considering seeing the model-based safety analysis preferably as deductive analysis and the classic FMEA further on as inductive analysis. Therefore, the systematic approach of consistent system engineering can again be applied from the vehicle level all the way down to the silicon structures and the software development.

6.4 Approvals/Releases

Releases are already addressed in ISO 9000 and therefore also in ISO TS 16949.

From *ISO 9001:2008*, Chap. 7.3.3 "*Development Results*":

The development results need to have a form, which is suitable for the verification against the development inputs and be approved before the release.

Also the term "Product approval" is used in different ways. This means that certain activities such as products need a release. ISO 9001 does not specify how such a release should take place. It is again the task of the respective management system to define the way in which such a release should be managed and what the subject of the particular release is.

Documented releases require from the decision making people that they are aware of the correctness and appropriateness of the relevant activities and the achieved characteristics and that they confirm that those have been fulfilled. This requires that the commissioned people have the sufficient competence for dedicated release. Negligently or grossly negligent performed releases can lead to damages or danger, which can cause the legislator or the insurance to get involved and become active. What legal consequences such decisions can have for the company or even for an individual person should not be discussed here. We should only bring attention to the further regulations and laws; especially, if we speak of releases of a clearly characterized safety activity or a safety relevant product.

6.4.1 Process Releases

In many APQP standards the product is released before the process. This is based on the assumption that when the product meets the requirements and targets, the process cannot have been completely wrong. If the process is released first, it is probably a production process release, which is then the prerequisite for a production of the product in line with the market requirements. Furthermore, it is assumed that if the process went well, the product will also prove to be of a certain quality. This in particular can lead to tremendous misconceptions.

This is why VDA suggests a process, product and project release (Fig. 6.6).

This figure shows that the respective maturity level dependency for multilevel supply chains.

Here we can see the work results and milestones for supply chains and the projects and products, which support this supply chain.

This milestone or maturity concept and its process protection are primarily used for the early detection of project risks, whereas safety issues of the product are one of such (Fig. 6.7).

The figure shows typical risk in the various phases and gives examples to minimize the risk for the product.

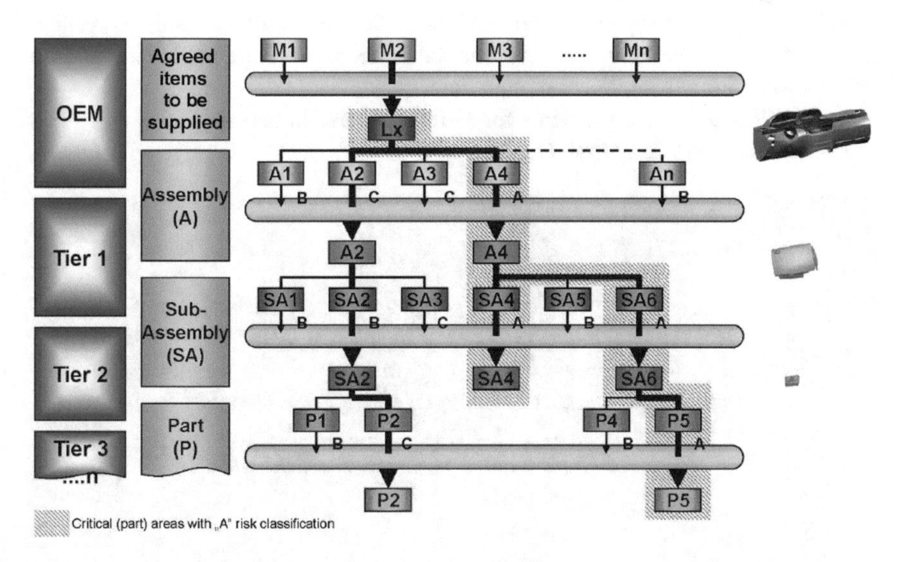

Fig. 6.6 Supplier Management—Specifiying the critical path source: VDA publication "maturity level assurance for new parts", 2nd edition, 2009

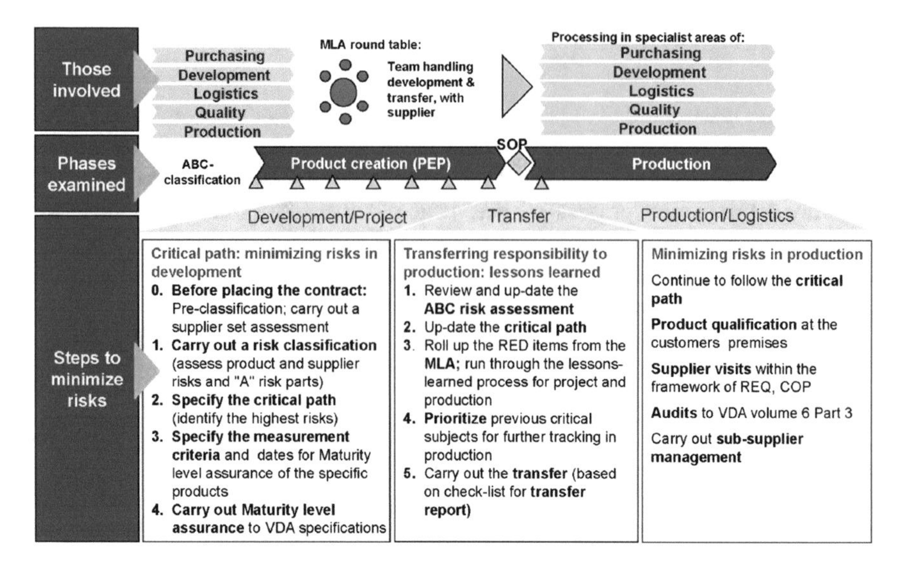

Fig. 6.7 VDA maturity model [3], maturity-hedge for new parts (*Source* VDA maturity for new parts)

6.4.2 Release for Series Production

According to all APQP or PPAP standards, the release for series production is also given to suppliers, through the vehicle manufacturer or the person responsible in the superimposing hierarchy of the supply chain. However, in all standards the vehicle manufacturer reserved the right to prove the correctness of the production and the product even for sub-suppliers of suppliers.

ISO 26262 define requirements for product release in part 4.

ISO 26262, Part 4, Clause 11:

11 Release for production

11.1 Objectives

11.1.1 The objective of this clause is to specify the release for production criteria at the completion of the item development. The release for production confirms that the item complies with the requirements for functional safety at the vehicle level

11.2 General

11.2.1 The release for production confirms that the item is ready for series-production and operation

11.2.2 The evidence of compliance with the prerequisites for serial production is provided by

> – *The completion of the verification and validation during the development at the hardware, software, system, item and vehicle level; and*
> – *The successful overall assessment of functional safety.*
>
> *11.2.3 This release documentation, forms a basis for the production of the components, systems or vehicles, and is signed by the person responsible of the release*

Particularly the last requirement stating that such a release needs to be signed by people is very common in the automobile industry. However, a product liability lawyer would not unreservedly recommend signing such a release.

6.4.3 Production Part Approval Process (PPAP)

One of the mayor processes for the acceptance of the product by OEMs and higher Tiers is the PPAP)

PPAP is a process defined in various APQP standards but also derived by nearly any vehicle manufacturer or higher Tier.

This process defines how the acceptance from on higher organization in the supply chain should be handled, so that at the end of the process a "release for series production" could be agreed (Fig. 6.8).

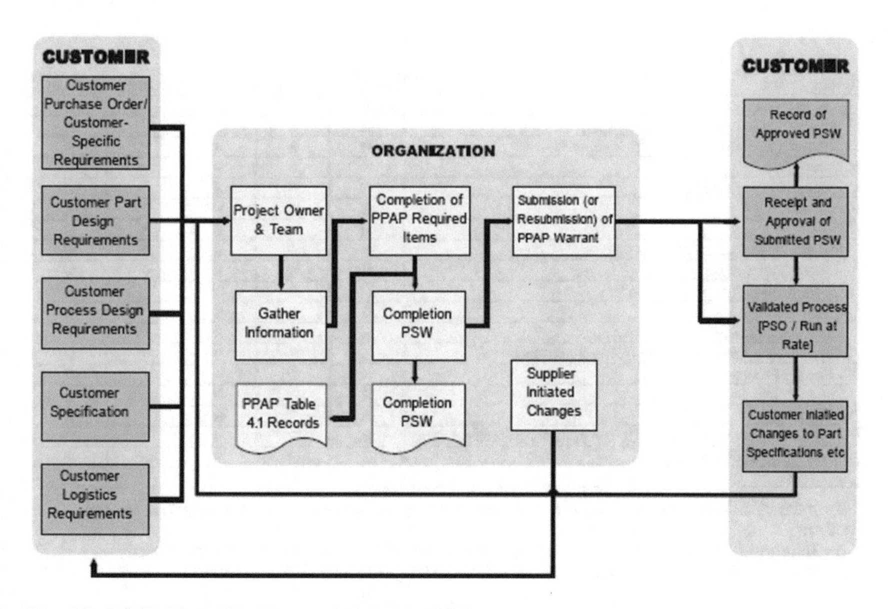

Fig. 6.8 PPAP Flow [2], (*Source* AIAG 4th. Edition)

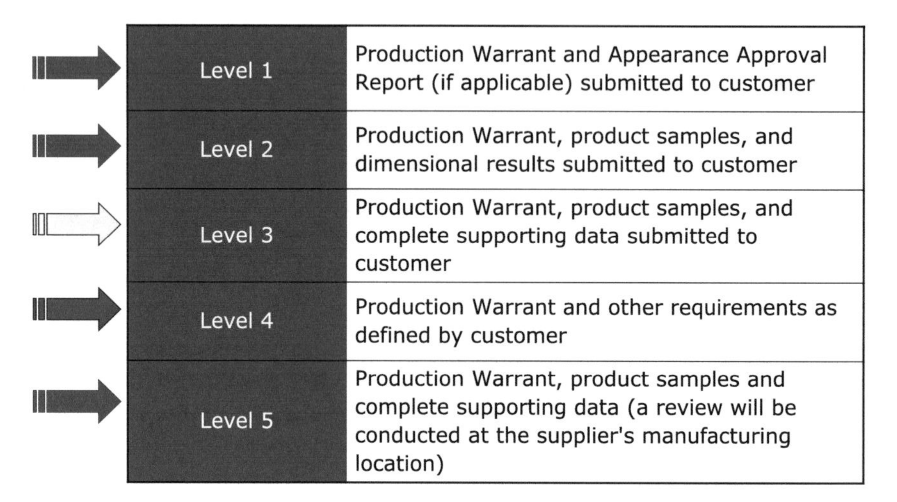

Fig. 6.9 PPAP Submission Levels (*Source* AIAG 4th. Edition)

There are 5 PPAP Levels which mainly lays down, what the customer expects when from his supplier. The supplier have to provide a warranty letter the PSW (product submission warrant), which is a declaration, that the delivered sample fulfils all agreed requirements, including safety requirements of course (Fig. 6.9).

Depending on the PPAP level the following work-products are expected (Fig. 6.10).

Requirement	Level 1	Level 2	Level 3	Level 4	Level 5
1.Design Record	R	S	S	*	R
2.Engineering Change Documents, if any	R	S	S	*	R
3.Customer Engineering approval, if required	R	R	S	*	R
4.Design FMEA	R	R	S	*	R
5.Process Flow Diagrams	R	R	S	*	R
6.Process FMEA	R	R	S	*	R
7.Control Plan	R	R	S	*	R
8.Measurement System Analysis studies	R	R	S	*	R
9.Dimensional Results	R	S	S	*	R
10.Material, Performance Test Results	R	S	S	*	R
11.Initial Process Studies	R	R	S	*	R
12.Qualified Laboratory Documentation	R	S	S	*	R
13.Appearance Approval Report (AAR), if applicable	S	S	S	*	R
14.Sample Product	R	S	S	*	R
15.Master Sample	R	R	R	*	R
16.Checking Aids	R	R	R	*	R
17.Records of Compliance With Customer Specific Requirements	R	R	S	*	R
18.Part Submission Warrant	S	S	S	S	R
19.Bulk Material Checklist	S	S	S	S	R

S = The organization shall submit to the customer and retain a copy of records or documentation items at appropriate locations

R = The organization shall retain at appropriate locations and make available to the customer upon request

* = The organization shall retain at the appropriate location and submit to the customer upon request

Fig. 6.10 Work-products as required by PPAP level (*Source* AIAG 4th. Edition)

The PPAP is strongly relying on production planning. The production sample is also one of the delivered work-products. The main engineering work-product is the Design-FMEA.

Many automotive companies expect beside the list of safety-related characteristics the entire safety case of the product as one of the additional required work-products, if the product is defined a safety-related product. Many companies require a safety case according to ISO 26262, so that all work-products have to be added to the list of work-products from the PPAP by the agreed work-products for a safety case according ISO 26262.

References

1. [ISO 26262]. ISO 26262 (2011): Road vehicles—Functional safety. International Organization for Standardization, Geneva, Switzerland.

ISO 26262, Part 4, Figure 2 217

ISO 26262, Part 4, Figure 3 218

ISO 26262, Part 4, Clause 7: 218

ISO 26262, Part 4, Clause 7.4.8.1: 219

ISO 26262, Part 8, Clause 9: 221

ISO 26262, Part 8, Clause 9: 223

ISO 26262, Part 8, Clause 9: 224

ISO 26262, Part 8, Clause 9: 225

ISO 26262, Part 8, Clause 9: 226

ISO 26262, Part 4, Clause 8: 231

ISO 26262, Part 3, Clause 8: 235

ISO 26262, Part 4, Clause 9: 236

ISO 26262, Part 4, Clause 9.2: 236

ISO 26262, Part 4, Clause 11: 246

2. [PPAP AIAG]. PPAP Production Part Approval Process AIAG 4th Edition, Automotive Industry Action Group, PPAP, 2006
3. [VDA maturity model]. Supplier Management—Specifiying the critical path, VDA publication "maturity level assurance for new parts", 2nd edition, 2009

Chapter 7
Confirmation of Functional Safety

Especially due to legal requirements, especially liability requires confirmation and a certain level of approval for the product under development are "State of the Art". In order to assure also for these confirmation sufficient or adequate transparence and traceability reports of the confirmations measures are required.

ISO 26262 [1] considers during concept and development phase 2 objectives.

ISO 26262, Part 2, Clause 6:

> *6 Safety management during the concept phase and the product development*
> *6.1 Objectives*
> *6.1.1 The first objective of this clause is to define the safety management roles and responsibilities, regarding the concept phase and the development phases in the safety lifecycle (see Figs. 1 and 2).*
> *6.1.2 The second objective of this clause is to define the requirements for the safety management during the concept phase and the development phases, including the planning and coordination of the safety activities, the progression of the safety lifecycle, the creation of the safety case, and the execution of the confirmation measures.*

The safety-lifecycle has been discussed detailed in the previous chapter of this book. For the second objective some further views based on the system engineering ideas have to be more detailed evaluated more detailed.

ISO 26262 provides three measures, which are necessary for the confirmation of functional safety for a product or an item

H.-L. Ross, *Functional Safety for Road Vehicles*,
DOI 10.1007/978-3-319-33361-8_7

ISO 26262, Part 2, Clause 6:

6.2 General
6.2.1 Safety management includes the responsibility to ensure that the confirmation measures are performed. Depending on the applicable ASIL, some confirmation measures require independence regarding resources, management and release authority (see 6.4.7).
6.2.2 Confirmation measures include confirmation reviews, functional safety audits and functional safety assessments:

- *the confirmation reviews are intended to check the compliance of selected work products (see Table 1) to the corresponding requirements of ISO 26262;*
- *a functional safety audit evaluates the implementation of the processes required for the functional safety activities;*
- *a functional safety assessment evaluates the functional safety achieved by the item.*

ISO 26262 uses the term "evaluate" in connection with the functional safety audit and assessment and "check" for the confirmation reviews. The assessment character for the functional safety audits could come from checking whether the safety activities have been implemented as planned on the basis of the functional safety concept.

The standard requires principally one Functional Safety Assessment, if functional safety has been achieved for an entire item on vehicle level in accordance with the given safety goals. Partial assessing of elements (systems, which do not build an item on vehicle level, components or electronic components, e.g. microcontroller) could also be performed within their system boundaries with regards to functional safety. However, the appropriateness or complete fulfillment of safety goals cannot be assessed for a specific vehicle.

Especially in distributed developments partial assessments are necessary, because the integrator on vehicle level is not able all necessary aspects of functional safety for complex products, such as software based multifunctional automotive systems.

Therefor ISO 26262 recommends a development accompanying Functional Safety Assessment and only the final evaluation of the product after integration, safety validation and other confirmation measures called a Functional Safety Assessment.

ISO 26262, Part 2, Clause 6:

6.2.3 In addition to the confirmation measures, verification reviews are performed. These reviews, which are required in other parts of ISO 26262, are intended to verify that the associated work products fulfil the project requirements, and the technical requirements with respect to use cases and failure modes
6.2.4 Table 1 lists the required confirmation measures. Annex D lists the reviews concerning verification and refers to the applicable parts of ISO 26262
6.2.5 Safety management includes the responsibility for the description and justification of any tailored safety activity (see 6.4.5).

Verifications during conception and product development are required as a mayor input for all confirmation measures. In ISO 26262, Part 2, annex D some proposals for planning and execution of verifications and other confirmation measures are shown.

For the confirmation measures, the standard provides some further detailed requirements.

ISO 26262, Part 2, Clause 6.4.7:

6.4.7 Confirmation measures: types, independency and authority
6.4.7.1 The confirmation measures specified in Table 1 shall be performed, in accordance with the required level of independency, Table 2, 6.4.3.5 i), 6.4.8 and 6.4.9
NOTE 1 The confirmation reviews are performed for those work products that are specified in Table 1 and required by the safety plan.
NOTE 2 A confirmation review includes the checking of correctness with respect to formality, contents, adequacy and completeness regarding the requirements of ISO 26262.
NOTE 3 Table 1 includes the confirmation measures. An overview of the verification reviews is given in Annex D.
NOTE 4 A report that is a result of a confirmation measure includes the name and revision number of the work products or process documents analysed [see ISO 26262-8:—, 10.4.5 (documentation)].
NOTE 5 If the item changes subsequent to the completion of confirmation reviews or functional safety assessments, then these will be repeated or supplemented [see ISO 26262-8:—, 8.4.5.2 (change management)].
NOTE 6 The aim of each confirmation measure is given in Annex C.
NOTE 7 Confirmation measures such as confirmation reviews and functional safety audits can be merged and combined with the functional safety assessment to support the handling of comparable variants of an item.

Confirmation measures	Degree of independency				Scope
	applies to ASIL				
	A	B	C	D	
Confirmation review of the hazard analysis and risk assessment of the item (see ISO 26262-3:—, Clause 5, ISO 26262-3:—, Clause 7 and if applicable, ISO 26262-8:—, Clause 5) -independence with regard to the developers of the item, project management and the authors of the work product	I3				The scope of this review shall include the correctness of the determined - ASILs, and - QM ratings of the identified hazards for the item, and a review of the safety goals
Confirmation review of the safety plan (see 6.5.1) - independence with regard to the developers of the item, project management and the authors of the work product	-	I1	I2	I3	Applies to the highest ASIL among the safety goals of the item
Confirmation review of the item integration and testing plan (see ISO 26262-4) -independence with regard to the developers of the item, project management and the authors of the work product	I0	I1	I2	I2	Applies to the highest ASIL among the safety goals of the item
Confirmation review of the validation plan (see ISO 26262-4) -independence with regard to the developers of the item, project management and the authors of the work product	I0	I1	I2	I2	Applies to the highest ASIL among the safety goals of the item
Confirmation review of the safety analyses (see ISO 26262-9:—, Clause 8) -independence with regard to the developers of the item, project management and the authors of the work products	I1	I1	I2	I3	Applies to the highest ASIL among the safety goals of the item
Confirmation review of the software tool qualification report[a] (see ISO 26262-8:—, Clause 11) -independence with regard to the persons performing the qualification of the software tool	-	I0	I1	I1	Applies to the highest ASIL of the requirements that can be violated by the use of the tool.
Confirmation review of the proven in use arguments (analysis, data and credit), of the candidates. See ISO 26262-8:—, Clause 14. -independence with regard to the author of the argument	I0	I1	I2	I3	Applies to the ASIL of the safety goal or requirement related to the considered behaviour, or function, of the candidate
Confirmation review of the completeness of the safety case (see 6.5.3) - independence with regard to the authors of the safety case	I0	I1	I2	I3	Applies to the highest ASIL among the safety goals of the item
Functional safety audit in accordance with 6.4.8 - independence with regard to the developers of the item and project management	-	I0	I2	I3	Applies to the highest ASIL among the safety goals of the item
Functional safety assessment in accordance with 6.4.9 -independence with regard to the developers of the item and project management	-	I0	I2	I3	Applies to the highest ASIL among the safety goals of the item

The notations: -, I0, I1, I2 and I3 are defined as:

-: no requirement and no recommendation for or against regarding this confirmation measure;

I0: the confirmation measure should be performed; however, if the confirmation measure is performed, it shall be performed by a different person;

I1: the confirmation measure shall be performed, by a different person;

I2: the confirmation measure shall be performed, by a person from a different team, i.e. not reporting to the same direct superior;

I3: the confirmation measure shall be performed, by a person from a different department or organization, i.e. independent from the department responsible for the considered work product(s) regarding management, resources and release authority.

a a software tool development is outside the item's safety lifecycle whereas the qualification of such a tool is an activity of the safety lifecycle

Fig. 7.1 Table 1: confirmation activities and their level of independency (*Source* ISO 26262, part 2, Table 1)

The three confirmation measures are further specified in Table 2 (Fig. 7.2).

ISO 26262 provides in Table 2 the mayor corner points of the 3 confirmation measures. In a footnote it indicates that the review and audit report can be included in the assessment report.

The tables (Fig. 7.1) also give an indication what level or **degree of independence** is required for what confirmation measure. The concern is, that people from their direct superior or other people in the hierarchy could be influenced to neglect important safety activities, or others violating an adequate result according to given safety requirements.

Generally, it will never be possible to describe, which safety activities are suitable for which risks; in any case it was not the aim of ISO 26262 to provide standardized concrete safety measures for certain failure scenarios. This is why it will also not be possible to say, which confirmation measures are necessary, adequate or suitable for which safety activities. The confirmation measures need to be planned based on the safety concept.

7.1 Confirmation Reviews

"Confirmation Review" is only mentioned in the tables of part 2. Apart from these tables there are no requirements for this kind of confirmation measure. But the objective of Confirmation Reviews is to build the bridge between other verifications as required or become necessary due to given requirements from the standard. The tables show that the key task should ensure the consistency and compliance of the norm according to ISO 26262 as well as that it is also a matter of standard compliance and work results and its documentation in defined work-products. However, most of the results also have to undergo verification according to ISO 26262, part 8, clause 9, which should show the consistency, correctness and completeness of essential work results. Essential however would be consistency checks of all work results, just like it is required later for the Safety Case. The tables label the confirmation reviews for the important work results:

- Hazard Analysis and Risk Assessment
- Project safety plan
- Vehicle system integration and test plans
- Validation plan
- Safety analyses
- Tool qualification
- Argumentation of "Proven-in-Use" (PIU)
- Completeness of the safety proof

Why the definition of the item, the functional safety concept, the component integration and their tests, the safety validation and the qualifications of hardware and software components do not undergo a confirmation review is unclear. However, some of these work results need to be verified.

Since the Confirmation Reviews are placed between the verifications and the Assessment of Functional Safety, it would be advisable to combine these activities as well as possible. The necessary independency has already been achieved for the verifications through a different person, all content, which requires specific qualified personnel, should be included through verifications. The insufficient independencies as required by the standard especially for higher ASIL could be complementary examined through the Confirmation Reviews. If a technically sufficient examination of consistency, completeness, transparency and correctness could be established through a useful combination of Confirmation Reviews and verifications, we could provide essential input for the Functional Safety Assessment by a multistage approach. It would be recommended to define such multi-stage approach along the maturity model of the given APQP standard. Because the maturity and specially the way what are the purpose of specific sample deliveries relies on implementation of safety measures in the process, safety mechanism in the sample and their verifications.

If Confirmation Review during all relevant steps during the activities according the safety lifecycle are planned, the final Functional Safety Assessment would be a respectively supplementing examination. It would complement the Safety Case by the safety validation and the assessment of the appropriateness of safety goals and their fulfillment for the confirmation of functional safety of the vehicle.

Verification of safety activities

ISO 26262 does not address many of requirements for the alignment of safety activities (processes) to the product development in general. Also, the Confirmation Measures in the standard are not described precisely so the verification of the safety activities, which can result from the interleaving of safety activities, has never been described. Only in ISO 26262, Part 8, clause 11 (Confidence in the use of software tools) some indications can be found that the process, which ISO 26262 is based on, should be safe by itself. Since the norm never described this process, the chains will unfortunately be destroyed in project planning. Therefor verifications of safety activities shouldn't be neglect.

Figure 7.3 as a sketch of the process structure in ISO 26262 runs through the entire requirements of the standard. It could be complemented with the problem solving and change process, the configurations and documentations management as well as the module management process. Basically, the verification is needed on all horizontal development layers. Above we can see the example for the system-components interface. However, the verification has to also take place between the functional and technical safety concept as well as the horizontal architecture interfaces in the components. Generally, each input and output of a safety activity should be verified. For multiple horizontal system levels the verification is called after each interface. The benefit is that the work results once again become input for the next phases after the verification, which ensures the confidence and resulting safety for the input and the predeceasing activities (Fig. 7.3).

This means that for example if the system architecture is verified including the allocated requirements and later in a further verification the system design is verified based on the previous requirements and architecture, an iteration cycle becomes transparent. This means, we prove the input of all activities against the developed output of the previous activities. The verification however, occurs parallel, so that requirements, architecture and design are continuously tested against the output of requirements, architecture and design of the previous phase. Therefore, each process error has to be revealed in the verification through the comparison of the output presented to the respective input. A basic requirement, not only for safety

Aspect	Review for confirmation	Audit of Functional Safety	Assessment Functional safety
The evaluation	Work result	Implementation of the processes that are required for functional safety.	Defined vehicle system, ISO 26262-3: 2011, Chapter 5
Result	Review Report (a)	Audit report (a) in accordance with ISO, Part 2, 6.4.8	Assessment Report ISO, Part 2, 6.4.9
Responsibility of the person performs the action	Evaluating the conformity of the results to the relevant requirements of ISO 26262	Assessment carried out required Processes	Evaluate the achieved functional safety. Creating a recommendation for acceptance, conditional acceptance or rejection in accordance with ISO, part 2-6.4.9.6
Time during the safety lifecycle	After the completion of the relevant Safety activities .. Completion Before the production release.	During implementation of the required processes.	Progressive during development or in a block. Completion Before the production release.
Scope and level of detail	According to the Safety plan	Carrying out the processes according to the defined activities, as referenced or specified in the safety plan.	The work results in accordance with the Safety plan, the required processes performed and the reviews of the Safety measures taken, which can be evaluated during the development of the vehicle system.
(A) This Reports in Assessment Report functional safety are introduced.			

Fig. 7.2 Table 2: confirmation measures and their characterization (*Source* inspired by *ISO 26262, part 2,* Table 2)

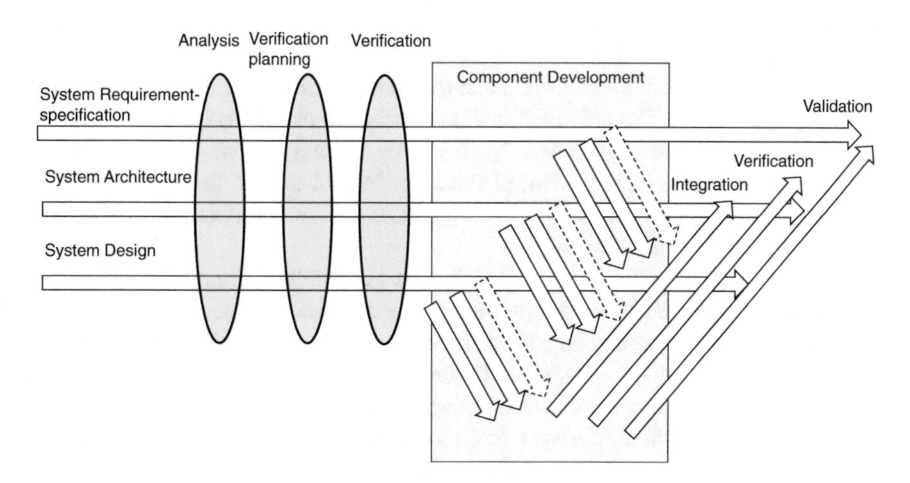

Fig. 7.3 Process verification model similar to ISO 26262

requirements, says that the output needs to be reproducibly generated based on the defined input. Since any verification needs to also determine the consistency, completeness and correctness, process failures can be detected through this test for the requirement, architecture and design development. If those process failures have not been considered for the verification planning, the statement that the process of

ISO 26262 is safe itself is untenable. Originally these aspects used to be described in the functional safety reviews. Unfortunately, after renaming it to 'confirmation reviews' in the later final forms, a lot of these aspects were lost. Since systematic failures, which are caused by process errors, tool impact errors or also human errors, can lead to inconsistencies, they should be detectable in well-planned verifications. Since the idea of the development processes derive from the production processes, we can also find good examples for the process verification. In production engineering such process verification is called a locking concept. It shows that even for an incorrect input, the production process is capable of detecting such failures through production monitoring. Also in this case the product is not changed through the verification. The change happens, for example, through follow-up treatment or because defective parts are sorted out. Technically, we can say that verifications and analyses (as special forms of verifications) are the essential initiators of change processes. The newly treated part however needs to once again go through the verification before it can be further processed. Production engineering says that the earlier an inconsistency is detected the cheaper is the follow-up treatment. This measure can also be very well applied for the development processes. ISO 26262 includes a chapter in part 8, which covers the tool qualification. However, as of late not many tools exist, which are appropriately qualified or if qualified used this method. Verifications should be planned accordingly.

If activities, which are supported by the tools, can emphasize safety relevant product influences verifications should lead to inconsistencies. Simply considering's, principals and many methods are based on process verifications even Process- and System-FMEAs are very similar. Considering a System-FMEA in a way that systematic failure can influence the function of a product, measures have to be taken against it. Possible malfunction (which are mainly systematic errors) in the Process-FMEA are controlled by measures during the production process, which is mainly the aim of a Control Plan. Analogical to that, the System-FMEA evaluates systematic errors during development process and determines adequate implemented safety mechanism.

If in fact, each possible systematic failure that can influence the failure behavior or important characteristics of the product and needs to be compensated with safety measures such as implemented safety mechanism. ISO 26262 did not distinguish between verifications of work-products and the process. The first verification, which should be strongly recommended, is the test of the intended functions, which are the basis for the item, the system on vehicle level. This verification indicates whether the functions can lead to hazards even if they functioning correctly. In this context we speak about the safety-of-use. Consequently, the Item Definition should be verified. If it turns out to be incorrect undetected inconsistency in the hazard and risk analysis should be expected.

It is important, that the planning of the analyses and verifications considers this and also effectively plans appropriate process locking similar to the production process also for the development process. This is especially evident for the planning of dissimilar or divers functions for example for an ASIL decomposition. If one algorithm is developed in Australia and one in Scandinavia it does not indicate whether the same systematic failures have been produced. However, if clearly different development targets are planned as process instructions, which cover the same safety goal, systematic failures can be detected by the safety technical inconsistencies. In this context for example one algorithm could be calculated with real numbers and the other with integer values or one function could be integrated through multiplication and the other addition like a Laplace transformation. There is also the possibility that in the product development, asymmetric conceptions can be planned for the test concepts, which then lead to the desired inconsistencies. Also, through failure injections or sample tests through multiple series, process failure tolerances can be tested for products, just like with the process capabilities for the production systems. Such dissimilar approaches need to be planned in a way that the system could only work, if the system is safe, any potential error would lead to a detected inconsistency which could be detected by an implemented safety mechanism. Of course such principles could be also implemented by fail-operational systems, but it requires an implementation of fully redundant signal chains and in case of inconsistencies a degraded function has to be still able to perform the safety-related intended function of the ITEM.

7.2 Functional Safety Audits

In this context the relation to CMMi appraisal and SPICE Assessment are often discussed, whereby the target of functional safety is not a matter of determining process improvement potential. The appropriateness of safety activities for their environment is seen as confirmation review, according to part 2, Table 2 (see Fig. 7.2). Whether the safety activities are appropriate related to given safety goals, is subject of the Assessment of Functional Safety. This also means that in order to tailor the safety lifecycle to the stipulated safety concept no SPICE assessor is needed but rather a safety expert. This does not mean that certain safety aspects cannot, should not or may not, be reasoned through the process. But the target of Functional Safety Audit is an evaluation of activities in line with the standard and adequate to realize the Functional Safety Concept. The degree of compliance to a given v-model is not important for functional safety. This is more often the case for the adjustment of activities because of certain tools or for streamlined processes for the application or development of variants of a useful approach. However, in order to do so a safety process (also defined in safety manuals, safety configuration handbooks or process manuals) should be accordingly planned in the concept phase. In order to plan such a process or the sequence of necessary safety activities, a safety specialist who often called a safety manager is required.

The safety process needs to be developed based on the item, the safety goals and the safety concept, since otherwise is it not possible to retrieve the necessary work-products for the safety case out of different activities. A project safety plan does not only describe the objectives of activities but also of the individual methods to be applied and intermediate targets such as the product or safety maturity in relation to the final product. For example, a fault tree analysis can be used for the identification of cut-sets, for the development of safety architecture or the identification of dependent failure. Even for the same product, fault trees analysis (FTA) can look very different, depending on which objectives and requirements of ISO 26262 should be met. This type of process analysis is based on a lot of individual activities, which have been derived from ISO/IEC12207 and became later process or process assessment models for SPICE or CMMi. The strategy or the objectives of planning of safety activities is to develop a sequence of activities that lead to a safety case based in a functional safety concept and given safety goals.

7.3 Assessment of Functional Safety

The Functional Safety Assessment is described in part 2 as part of Confirmation Measures and in part 4 of ISO 26262 at the end of system development after the safety validation and before the release for serial production for the product to be developed. Primarily object of the Functional Safety Assessment is the assessment of the Safety Case according to the requirements in part 2 of ISO 26262. The requirements of how the Functional Safety Assessment should be performed are given in part 4 of ISO 26262.

Part 4, clause 10 mentions the following objectives and requirements for the Assessment of Functional Safety:

ISO 26262, Part 4, clause 10:

10 Functional safety assessment
10.1 Objectives
The objective of the requirements in this clause is to assess the functional safety that is achieved by the item.
10.2 General
The organisational entity with responsibility for functional safety (e.g. the vehicle manufacturer or the supplier, if the latter is responsible for functional safety) initiates an assessment of functional safety.
10.3 Inputs to this clause
10.3.1 Prerequisites
The following information shall be available:

- *safety case in accordance with ISO 26262-2: —, 6.5.3;*
- *safety plan (refined) in accordance with 5.5.2, ISO 26262-5: -, 5.5.2 and ISO 26262-6:-, 5.5.2;*
- *confirmation review reports in accordance with ISO-26262-2: —, 6.5.4;*
- *audit report if available in accordance with ISO-26262-2: —, 6.5.4; and*
- *functional safety assessment plan (refined) in accordance with 5.5.4.*

Surprisingly the standard does not require the product itself as an input for the activity, which does not mean that the Functional Safety Assessment could be simply reduced to a document check. A couple of further requirements, which result from the verification and validation activities and their requirements, disagree. Also the required assessment plan in Chap. 5.5.5, which results from the single requirement that the Assessment of Function Safety needs to be planned at the beginning of system development, indicates that the entire product development process needs to be assessed so that proven evidence for the Functional Safety Assessment could be provided.

Furthermore, the Safety Case is an essential input for the assessment and thus the safety validation is also already considered as an essential input for the assessment.

ISO 26262, Part 4, clause 10.4:

10.4 Requirements and recommendation
10.4.1 This requirement applies to ASILs (B), C, and D, in accordance with 4.3: For each step of the safety lifecycle in ISO 26262-2: —, Fig. 2, the specific topics to be addressed by the functional safety assessment shall be identified.
10.4.2 This requirement applies to ASILs (B), C, and D, in accordance with 4.3: The functional safety assessment shall be conducted in accordance with ISO 26262-2: —, Clause 6.4.9 (Functional safety assessment).

Basically, those two requirements say that the entire safety lifecycle needs to be considered for the Functional Safety Assessment. This explicitly includes that the correct planning of safety activities (tailoring of safety lifecycles) influences the assessment. Furthermore, a direct assessment of function safety is only recommended for ASIL B and required for ASIL C and D. This is only the view of the norm and also only based on the requirements, which the standard general requires for Functional Safety Assessments. Since the safety validation and the necessary verifications need to be conducted for all ASIL in any case, it is only advisable to find a useful solution within the relevant organization. An assessment of safety relevant products however still needs to be performed because of product liability reasons. But the assessment needs to be not in line with the standard. Which adjustments are necessary in the individual case can only be determined by the respective organization.

7.4 Safety Case

The target of the safety case is to provide arguments for the safety of the item, or the system on vehicle level. According to ISO 26262 the given requirements from the standard address only the functional safety of the EE-system. Other technology or

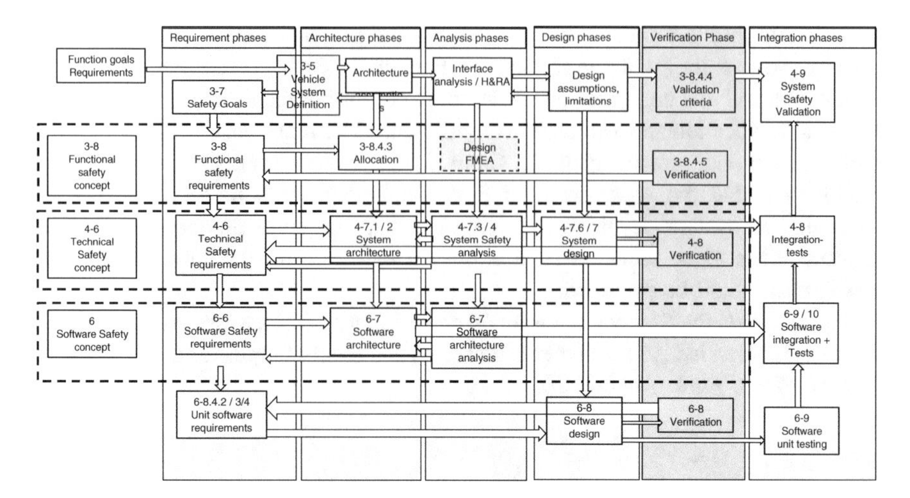

Fig. 7.4 Safety Case as argument for the "Confirmation of Functional" Safety based on the validated Safety Goals and verified work-products as planned based on the Safety Concept

external measures are also addressed, but impacts from outside of the boundary or malfunctions affected by external systems or even realized in other technology are not direct addressed. Especially touch protection (electrical safety), chemical or toxically effects due substances are not addressed. Especially for people protection as required in the machinery directives, no information could be found in ISO 26262. This means that safety case according ISO 26262 is only a matter of safety technical correct functions and their deterministic behavior in case of failure (Fig. 7.4).

ISO 26262 requires mainly the compiled work-products derived from the safety activities during concept and development phase as planned based of the safety goals and the defined safety concept. Those should provide sufficient evidence for the functional safety of the item.

The safety case based on the safety argumentation of the following aspects:

- Are the scope and work-products of the individual safety activities consistent?
- Were the failure and safety analyses sufficiently and correctly performed?
- Were relevant adequate safety measures implemented for the imaginable malfunctions?
- Verification of all relevant work results
- Validation of safety goals (are they correct, sufficient and fully achieved)
- Assessment of all activities and work-products included in the safety case

The chapter for the Safety Case has been intentionally placed at the end of this book since the reproducible evidence of functional safety for an Item represents the aim of ISO 26262. Safety Validation and Functional Safety Assessment provide mayor arguments for the Safety Case although the entire Safety Case is object of the Functional Safety Assessment. The safety concepts of today's vehicle systems are simply too different to describe a static process. As previously mentioned at the beginning, it has never been the aim of ISO 26262 to be a guideline for the safe development of vehicles. The aim has always been to provide indications in the form of requirements, which need to be considered for the safe development of vehicles. It remains unclear whether meeting all the requirements, results in creating a functionally safe system—everyone who has read this book should draw their own conclusion.

References

1. [ISO 26262]. ISO 26262 (2011): Road vehicles – Functional safety. International Organization for Standardization, Geneva, Switzerland.

Index

H.-L. Ross, *Functional Safety for Road Vehicles*,
DOI 10.1007/978-3-319-33361-8

Printed in the United States
By Bookmasters